Commercialization of Smallholder Horticultural Farming in Kenya

DEVELOPMENT ECONOMICS AND POLICY

Series edited by Franz Heidhues, Joachim von Braun and Manfred Zeller

Vol. 72

PL ACADEMIC RESEARCH

Commercialization of Smallholder Horticultural Farming in Kenya

Poverty, Gender, and Institutional Arrangements

Beatrice Wambui Muriithi

PL ACADEMIC RESEARCH

Bibliographic Information published by the Deutsche Nationalbibliothek
The Deutsche Nationalbibliothek lists this publication in the Deutsche Nationalbibliografie; detailed bibliographic data is available in the internet at http://dnb.d-nb.de.

Zugl.: Bonn, Univ., Diss. 2013

Library of Congress Cataloging-in-Publication Data
Muriithi, Beatrice Wambui, 1979-
Commercialization of smallholder horticultural farming in Kenya : poverty, gender, and institutional arrangements / Beatrice Wambui Muriithi. – Edition 1.
pages cm. – (Development economics and policy, ISSN 0948-1338 ; vol. 72)
Includes bibliographical references.
ISBN 978-3-631-65337-1
1. Farm produce–Kenya–Marketing. 2. Horticultural crops–Kenya–Marketing. 3. Farms, Small–Kenya. 4. Kenya–Commerce. I. Title.
II. Series: Development economics and policy ; Bd. 72.
HD9017.K42M87 2014
338.1'75096762–dc23

2014012381

Printed with financial support of the
German Academic Exchange Service (DAAD).

Cover photo: © Beatrice Wambui Muriithi

D 98
ISSN 0948-1338
ISBN 978-3-631-65337-1 (Print)
E-ISBN 978-3-653-04518-5 (E-Book)
DOI 10.3726/978-3-653-04518-5

PL Academic Research is an Imprint of Peter Lang GmbH.
Peter Lang – Frankfurt am Main · Bern · Bruxelles · New York · Oxford · Warszawa · Wien

This publication has been peer reviewed.

www.peterlang.com

Acknowledgements

This work would not have been possible without the assistance of a number of people and institutions. First, I am thankful to God for giving me the grace to bring this work to fruition.

I am greatly indebted to my first supervisor Prof. Dr. Joachim von Braun, who provided guidance, encouragement, and constructive criticism in the friendliest way, in addition to offering moral and intellectual support. I am also grateful to Prof. von Braun for his kindness and concern for my well-being. I also wish to express my gratitude to my second supervisor Prof. Dr. George Noga who took time to read my work and gave timely feedback. I am very grateful to my tutor Dr. Julia Anna Matz for her invaluable comments on several drafts of this thesis. I am thankful for her timely and constructive feedback on my questions and writing, which made the completion of this work possible.

This study was accomplished with financial support from the German Academic Exchange Service (DAAD), without which it would not have been possible. I am equally thankful to the Fiat Panis Foundation for their additional financial support, which enabled me to carry out such an elaborate research effort. I am also grateful to the International Centre of Insect Physiology and Ecology (ICIPE) for allowing me to use their data. I am particularly grateful to the director, Prof. Dr. Christian Borgemeister for hosting me at his institute during my fieldwork and to Prof. Dr. Dagmar Mithoefer for taking time to discuss the secondary data and my field preparations. I also wish to thank my enumerators whose kind cooperation during data collection was invaluable to the completion of this study. I also appreciate the cooperation of the Kenyan Ministry of Agriculture, especially the front-line extension workers, vegetable export companies, and farmers who sacrificed their precious time to participate in our interviews.

I am grateful for the unique atmosphere of support and collaboration that I received at the Centre for Development Research (ZEF). I particularly want to thank Dr. Guenther Manske, Mrs Rosemary Zabel, and Mrs Maike Retat-Amin for the friendship and many administrative activities that they undertook on my behalf. I am grateful for their outstanding efficiency in responding to my concerns, which has contributed greatly to the successful completion of my studies. I am also thankful to all of my ZEFb colleagues for their moral and professional support, and for their encouragement during my studies.

I would like to thank my dear husband, George Kihuria, and our son, Collins Muthuki, for the moral support and patience with which they handled my many days, months, and years away from home. Their love and tolerance encouraged me to complete my studies as soon as possible. I also owe special thanks to my parents for their great inspiration during my academic career. My late father's encouragement to successfully complete my studies is still fresh in my memory and challenges me to live up to his dreams. Without my mother's sacrifices during my early school years, I would not be where I am today. Finally, thanks to all of my siblings and many friends, including my friends in Germany, the East African community in Bonn, and friends from the American Protestant Church, who offered me their moral support and made my stay in this beautiful country a fantastic experience.

Table of Contents

List of Tables

List of Figures

Acronyms and abbreviations

AGRA	Alliance for Green Revolution in Africa
ASALS	Arid and Semi-Arid Lands
ASDS	Agricultural Sector Development Strategy (2010–2020)
CAADP	Comprehensive Africa Agriculture Development Programme
CBA	Cost-benefit analysis
EAC	East African Community
ERS	Economic Recovery Strategy
EU	European Union
FFV	Fresh Fruits and Vegetables
FPEAK	Fresh Produce Exporters Association of Kenya
GDP	Gross Domestic Product
AGRA	Alliance for Green Revolution in Africa
GlobalGap	Global Good Agriculture Practices
GoK	Government of Kenya
HACCP	Hazard Analysis and Critical control points systems
HCDA	Horticultural Crops Development Authority
ICIPE	International Centre of Insect Physiology and Ecology
IFAD	International Fund for Agriculture Development
IFPRI	International Food Policy Research Institute
KARI	Kenya Agricultural Research Institute
KHCP	Kenya Horticultural Competitiveness Project
KHDP	Kenya Horticulture Development Programme
KEPHIS	Kenya Plant Health Inspectorate Services
KenyaGap	Kenya Good Agriculture Practices
KIPPRA	Kenya Institute of Public Policy Research and Analysis
KNBS	Kenya National Bureau of Statistics
KRA	Kenya Revenue Authority
Ksh	Kenyan Shillings
MDGs	Millennium Development Goals
MoA	Ministry of Agriculture
NGO	Non-Governmental Organization
PMOs	Production and Marketing organizations
SRA	Strategy for Revitalizing Agriculture

UK	United Kingdom
UN	United Nations
USAID	United States Agency for International Development
WHO	World Health Organization

Chapter One

1. General Introduction

1.1 Background and research problem

Agriculture and agribusiness remain fundamental instruments for sustainable growth and poverty reduction in many developing economies. In agriculture-based economies such as in Sub-Saharan Africa, agriculture generates 29% of the Gross Domestic Product (GDP) on average and employs 65% of the labour force (World Bank, 2007). Together, agriculture and agribusiness account for nearly half of the GDP in Africa (World Bank, 2013). Today, about 1.4 billion people live on less than USD $1.25 a day (UN, 2012). The majority of these poor are found in countries with agriculture-based economies. Over 75% of them depend on agriculture for their livelihood and live in rural areas. Promoting agriculture therefore remains imperative to meeting the MDG goal of halving poverty and hunger. Raising agricultural productivity is particularly crucial if agriculture is used as a basis for economic growth (World Bank, 2013). Growth in agricultural productivity can be achieved by improving the competitiveness of agricultural products in international, regional and domestic markets; especially for high-value horticultural crops. In the recent past, the intensification and commercialization of smallholder agriculture in Africa has been identified by many government and development agencies as a key means for delivering accelerated growth in agriculture, which is critical to meeting MDGs in Africa (Wellard, 2011). The comparative advantage of the commercialization of smallholder farmers over subsistence production therefore presents a strong justification for its promotion while seeking equitable distribution of benefits and costs accrued from this process.

Past studies show the existence of high demand for commercialization of horticultural farming (Nyoro and Jayne, 2005; Virchow, 2008). The driving forces include globalization, the expansion of modern marketing channels (for example supermarkets), rural to urban migration, increased consumer food safety and quality awareness, and population growth (Muendo & Tschirley, 2004; Neven & Reardon, 2004; Jaffee, 2005; Adekunle et al., 2012). At the same time, the commercialization of smallholder agriculture in developing markets faces numerous challenges. These include information barriers, high transportation costs (attributed to physical infrastructure), and high market fees. In addition, rapidly changing markets create new challenges for the sub-sector. Already stringent sanitary and phytosanitary standards in the global market have been tightened while new standards are being applied to address previously unknown and unregulated risks, especially in high-value markets. Meeting those standards represents potential barriers, especially to small producers seeking to expand their trade opportunities in the international markets. Policy interventions therefore require an understanding of how smallholders are affected by these challenges and how institutional arrangements can enhance opportunities. Chapter Two of this dissertation presents the results of an evaluation of the dynamics of smallholder participation in commercialization in the context of emerging opportunities and challenges that are intended to improve this understanding.

In Kenya, like in many other developing countries, the horticultural sub-sector is often considered a key element of pro-poor development strategies due to its direct link to rural economies (intensive use of land and unskilled labour) (Aksoy, 2005). This sub-sector is often the basis of livelihoods for a significant share of the population. In addition to income, small scale producers typically also have access to fresh vegetables and fruits that provide essential components of a healthy diet. The horticultural sector contributes a significant portion of the agricultural GDP, both from foreign exchange earnings and domestic sales. Various studies of the commercialization of horticulture demonstrate that household income increases upon the allocation of resources (labour, land, capital and other resources to production) to market-oriented horticulture, both in the product and labour markets. For example, a study in Kenya by McCulloch and Ota (2002) found significant positive impacts on poverty that the export horticultural industry had on households involved either as producers or labourers in this sub-sector. Positive income effects of the commercialization of high-value crops have also been reported in other African countries such as Senegal (Maertens et al., 2008), Ghana (Afari-Sefa, 2007), and Zimbabwe (Henson et al., 2005). Similar findings have been reported from other continents (for details see Tschirley, 2007). Although these studies have contributed important insights regarding the contribution of the commercialization of horticultural farming by smallholders, the welfare effects reported are mainly based on cross-sectional data, reporting income at the time of survey. A few exceptions include Carletto et al. (2009) and von Braun and Immink (1994) in Guatemala. However, similar studies are missing for Kenya. Furthermore, most of these studies qualify the commercialization of horticultural farming by looking at poverty effects from an income perspective. The contribution of commercialization of smallholder horticultural farming on household asset accumulation is missing. This is addressed in material presented in Chapter Three of this study.

Gender equality is crucial for agricultural development and for attainment of the MDGs (World Bank et al., 2009). Women play a critical and potentially transforming role in agricultural growth in the agriculture-based economies of developing countries, but they often face persistent constraints limiting their further inclusion in agriculture (IFPRI, 2012). According to the World Bank, women represent about 43% of the global agricultural labour force (World Bank, 2012). In Kenya, women account for over two-thirds of the total labour force in both the production and processing of horticultural crops (GoK, 2012). Gender effects of the commercialization of horticulture have been alluded to in past studies. However, most of these studies are skewed to the labour market (Barrientos et al., 2005; Tallontire et al., 2005; Maertens, 2009). Past studies have largely ignored gender effects of the commercialization of horticulture from a product market perspective with few exceptions (e.g. Dolan, 2001; Fischer & Qaim, 2012a). However, all of these studies agree that the gender effects of commercialization depend on who controls the income generated. Gains for women from the product market are frequently constrained by a lack of property rights to land resources. Land ownership is an important element of successful horticultural farming, especially where contract agreements are involved. The commercialization of horticulture could therefore contribute to gender inequality, thus affecting livelihoods of rural households. Chapter Four of this dissertation includes the results of an exploration of gender effects of the commercialization of horticulture from a product market perspective intended to contribute to the limited studies on this phenomenon.

Participation by Kenyan smallholder producers in high-value export markets has been declining (Ashraf et al., 2009; Narrod et al., 2009). Agricultural market support institutions and market arrangements such as contracts and farmer production and marketing organizations (PMOs) have been developed to overcome some of these barriers to participation. For example, collective action through PMOs has been observed by recent studies as a strategy for facilitating smallholder horticultural exporters to comply with international standards (Okello & Swinton, 2007; Narrod et al., 2009). Producer contracts on the other hand, facilitate access to farm inputs (improved variety seeds, fertilisers and pesticides), credit, technical advice (e.g. for the application of chemicals), and the produce market. Past studies have analysed the impacts of these agricultural market institutions on household welfare but have not quantified transaction costs involved in implementing and sustaining these institutions. The results of my assessment of these costs are presented in Chapter Five.

The premise of this study is that for an agriculture-dependent economy, participation in commercialization by rural smallholder farmers is indispensable. The identification of commercialization pathways, the determinants of choice of these pathways, their effects on rural livelihoods, and the evaluation of the performance of institutions that support smallholders to commercialize are thus of policy relevance in a developing country like Kenya. The specific objectives of this study are as follows:

1. To investigate the market pathways for commercialization of smallholder horticultural farming and to assess the factors that influence their participation along different pathways.
2. To assess the impact of horticultural commercialization among smallholders on household income and poverty.
3. To identify gender roles within the commercialization of horticultural farming and to assess their implications for household welfare outcomes.
4. To identify and evaluate the performance of institutions and institutional arrangements that promote the commercialization of smallholder horticultural farming.

Objective one, which is presented in Chapter Two aims at characterizing commercialization pathways of smallholder vegetable producers. The determinants of choice and the extent of the commercialization of these pathways are also assessed. In Chapter Three, the marginal effect of the commercialization of vegetables on household income and poverty is assessed. In Chapter Four, the analysis of the effects of commercialization of smallholder horticulture is extended by looking at gender roles in vegetable production and marketing and the implications of female participation in the commercialization of vegetables on household welfare outcomes. Finally, in Chapter Five, the institutions and institutional arrangements that promote the commercialization of smallholder horticultural farming are identified and their performance evaluated. A concluding chapter, which presents the synopsis of the study, acknowledges the limitations of this work, and provides suggestions for further research follows after Chapter Five.

1.2 Relevance of the horticultural sub-sector in the Kenyan economy

Horticulture is an important sub-sector of the Kenyan economy. It contributes to food security, poverty reduction, income and employment creation, foreign exchange, and raw material for agro-industries (GoK, 2012). The Kenyan horticultural sub-sector is comprised of vegetables, fruits, herbs, spices, and floriculture. Of the total value of horticultural production, vegetables account for about 44.6%, fruits for 29.6%, flowers for 20.3%, while the remainder is derived from nuts, and medicinal and aromatic plants (GoK, 2012). Smallholder farmers predominantly produce vegetables, thus this study focuses primarily on vegetable production. About 90% of the vegetables produced are consumed in the domestic market (Tschirley & Ayieko, 2008), which also accounts for 97% of fruits, and 87% of herbs and spices (HCDA, 2009; KHCP, 2011). Floriculture is dominated by large-scale producers, including foreign owned agribusinesses, and about 90% is exported, primarily to the EU market (HCDA, 2009). Overall, 96% of the total horticultural production is consumed domestically, while the remainder is exported. Despite its overall minority share of production the export sub-sector earns the country vast amounts of foreign exchange (GoK, 2012).

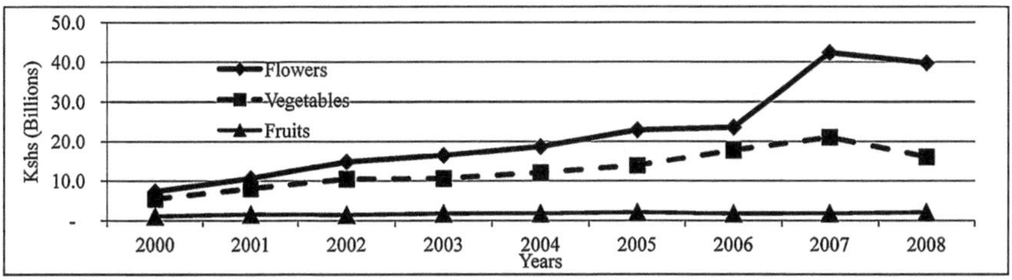

Source: Based on HCDA export statistics (2000–2010)

Fig. 1.1: Trends of Kenyan horticultural exports

This subsector contributes about 36% of the agricultural GDP and employs over six million people (GoK, 2012). Over 75% of these employees are women, who primarily work on farms and in processing (Barrientos et al., 2003). Fresh fruits and vegetables also provide essential vitamins and other fundamental micro-nutrients to low-income household diets. The horticultural sub-sector is the fastest growing agricultural sub-sector in Kenya and ranks second after tea in terms of foreign exchange earnings (HCDA, 2009). It contributes about 38% of national agricultural export earnings (GoK, 2010a). Between 1995 and 2009 the volume of produce exported increased by 35% for fruits, 60% for vegetables, and 66% for flowers (HCDA export statistics 1995–2009). Figure 1.1 presents the trends in major Kenyan horticultural exports over the last decade.

Kenya's horticulture exports have often been regarded a success story owing to their competitiveness in international markets, especially compared to exports from other Sub-Saharan countries. In the early 2000s, Kenya was ranked as the second largest horticultural exporter in Sub-Saharan Africa (after South Africa). It was also ranked the second-largest developing country exporter of flowers in the world after Colombia, as well as the second-largest developing country supplier of vegetables to the EU after Morocco (English et al., 2004). The success of this subsector has been attributed to a number of factors (McCulloch & Ota, 2002). Kenya's climatic and geophysical conditions in combination with its location on the equator allow year-round agricultural production, and provide the country with a competitive edge against competitors such as Egypt and Morocco. The use of relatively low-skilled labour for the production of horticultural crops, which are land and labour intensive, has also been cited in previous studies as providing Kenya and other Sub-Saharan countries with competitive advantage over regions such as Latin America (McCulloch & Ota, 2002). This may have changed in recent years, as the costs of unskilled labour have increased due to inflation. The success of Kenyan horticulture sub-sector can also be partly attributed to the minimal role that the government takes in it (Minot & Ngigi, 2004).

1.3 The role of small-scale producers in the horticultural sub-sector

It is estimated that over 90% of rural households produce some fruits and/or vegetables, either for home use or for sale. About 76% of these producers sell their produce in either the domestic or export markets (Tschirley et al., 2004). Muendo & Tschirley (2004) estimate that between 1997 and 2001, out of the total amount of vegetables produced vegetables, about 36% were retained on-farm for subsistence consumption, 52% were sold domestically, and about 12% were exported. There are various actors in Kenya that channel fresh fruits and vegetable to both the export and domestic markets. These include large- and, small-scale farmers, middlemen, export companies, and both local and international supermarkets.

Small-scale producers generally dominate the domestic market channels. Figures on the number of those involved in the export of fresh fruits and vegetables vary, based on the source and year cited. Estimates from the 1990s indicate that approximately 75% of the total value of vegetables exported was supplied by smallholders (Kimenye, 1993; Jaffee & Morton, 1995). In early 2000 the Horticultural Crops Development Authority (HCDA) estimated that smallholders produced 40% and 70% of exported fruits and vegetables respectively (Harris et al., 2001). Based on interviews with several exporters, Jaffee (2003) estimated that smallholders accounted for 27% of exported fresh vegetables and 85% of exported fresh fruits. This accounts for a combined average of 47% of the total fresh fruits and vegetables exported. The figure decreased from 47% in 2002 to approximately 30% in 2003 (Jaffee 2003; Minot & Ngigi, 2004). National estimates of the number of smallholders producing fruits and vegetable for export markets range from 13,000–16,000 (Jaffee & Morton, 1995), to 80,000 (Karuga & Masabayi, 2004). Estimates by Mithofer et al. (2008) in nine major export market vegetable producing districts indicate that approximately 12,000 smallholders were involved in vegetable production for export. Recent discussions between the author and experts in Kenya suggest that the number of small-scale producers has continued to decline. The experts attribute the decreased export market participation of smallholders to several factors, which I have explored in this study.

1.4 Market pathways for commercialization of vegetables in Kenya

Commercialization participation by smallholder horticultural farmers in Kenya can be broadly categorized into product (production of vegetables) and labour (on vegetable farms and processors) market pathways. As stated earlier, the highest proportion of marketed fresh fruits and vegetables is sold and consumed within the domestic market. The domestic horticultural market is four to five times larger by value than the export market (Tschirley et al., 2004). The domestic market supply chain can broadly be sub-divided into two parts; the "modern" market (mainly domestic supermarkets) and the "traditional" domestic market. According to Neven and Reardon (2004), domestic supermarkets impose stringent requirements on produce quality. They demand specific standards for size and appearance, and pay a premium for low maximum pesticide residues. Smallholders that are not organized face difficulties in meeting these requirements (Ngugi et al., 2006). In 2004, Neven and Reardon (2004) found that the largest percentage of fresh vegetables supplied to supermarkets were procured from brokers (45%), followed by medium-sized farms (about 25%), large-scale farms (about 15%), small-scale farms (10%), and the rest was imported. Tschirley et al. (2004) estimated that demand by domestic supermarkets would grow by about 10% by 2014, while Traill (2006) estimated a 16% increase in the share of total vegetables demanded by supermarkets by 2013. Rao and Qaim (2011) provide more recent insights on modern domestic channels. Although the authors did not estimate the growth of smallholder participation in domestic supermarkets, they found that participation in supermarket channels was associated with gains of up to 48% in average household income. Thus, this market channel is attractive to smallholders. The traditional marketing channels include wet markets, kiosks, groceries, and street vendors. These channels absorb the largest portion of the horticultural produce but have the lowest returns since they are not efficiently organized in comparison to supermarket supply chains (Tschirley and Ayieko 2008). While most horticultural income is derived from the domestic markets, there are very few studies focusing on domestic market channels. The export markets have attracted the most attention. Researchers and investors have neglected the domestic horticultural sub-sector for many years (World Bank, 2007 [cited in Tschirley, 2007]; Lenné and Ward, 2011). Figure 1.2 shows production volumes of vegetables sold in domestic markets. In addition to the vegetables shown in Figure 1.2, Irish potatoes occupy the highest position in terms of production volumes, with an average of 2.6 million tonnes per year in 2005, 2006, and 2007.

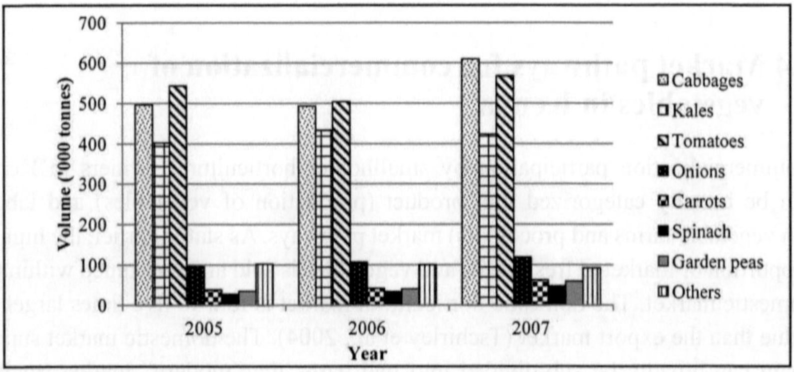

Source: HCDA (2008)

Fig. 1.2: Kenyan vegetable production for domestic market by crop and year

Regional markets are also important to the horticultural sub-sector in Kenya. Regional trade between Kenya and its immediate neighbouring countries (Uganda and Tanzania) takes place within the framework of the Custom Union (CU) established by the East African Community (EAC). Regional horticultural trade also involves Sudan, Somalia, South Africa, and Egypt. A recent study by USAID (2011) found that between 2006 and 2010, Kenya exported USD $428 million (six times the value of imports) of horticultural products to African countries. These products included processed and fresh fruits and vegetables, spices, flowers and miraa (khat). The annual growth rate of exports, however, was estimated to be smaller (14.3%) than that of imports (26.2%). In terms of income generation for smallholders, the products with the greatest potential for growth in African markets include potatoes, high-value vegetables, mango, and passion fruit (USAID, 2011). The main market destinations for Kenyan horticultural products during the five years covered by the USAID (2011) study were Uganda (10% of total regional exports), Tanzania (8%), Somalia (6.5% excluding miraa), and South Africa (5%). Processed fruits had the highest volume contributing about 37.5% of total exports, followed by fresh vegetables (17.9%), and fresh fruits (16.5%). Figure 1.3 presents Kenya's fresh vegetable exports to the East African regional market.

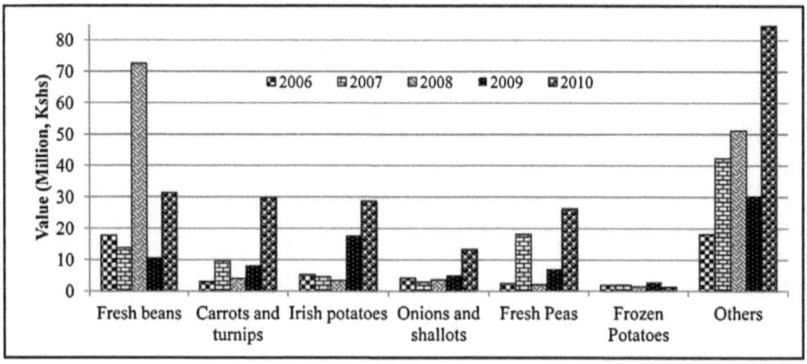

Source: KHCP (2011)

Fig. 1.3: Kenyan exports of fresh vegetable products to the regional market

Kenya's horticulture export industry depends heavily on European markets (80%) (GoK, 2012). Between 2001 and 2010, horticultural exports rose at an average annual rate of 15.7% (GoK, 2012). Export vegetables contributed about 24% of the total value of horticultural exports in 2011. The major vegetables exported to international markets include Asian vegetables, French beans, peas, carrots and turnips (HCDA, 2009). Figure 1.4 depicts trends in Kenyan vegetable production for export markets.

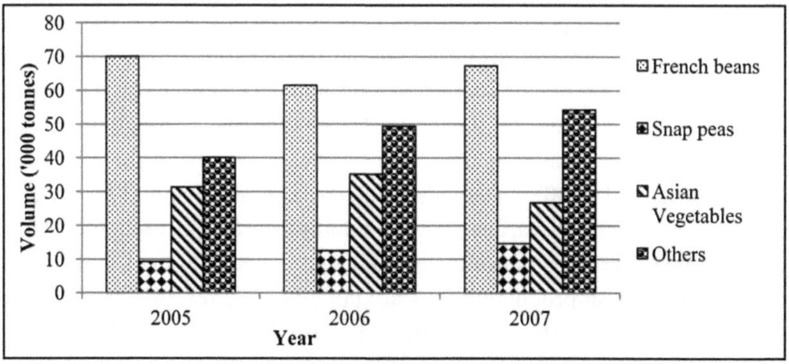

Source: HCDA (2009)

Fig. 1.4: Kenyan vegetable production for export markets by crop and year

The heavy reliance on European markets implies that any disturbances in these markets would have significant consequences for the Kenyan horticultural export industry. The leading European destinations for Kenyan vegetables are the United Kingdom (UK), France, Germany and the Netherlands. In 2009 the UK imported 54% of all Kenyan vegetable exports, followed by France (15%), and the Netherlands (11%) (HCDA, 2009). Kenya also exports to the Middle East and a few other European countries. Smallholders that produce for export markets operate either as individuals or as members of out-grower schemes, mostly on the basis of contractual arrangements. This is facilitated by collective action through PMOs. This is necessary in order pool sufficient volumes of produce to attain economies of scale and to invest cooperatively in facilities required to meet production standards (Paalhaar & Jansen, 2011). The success of these associations in linking smallholders to export markets is mixed. Their success (or failure) is often associated with structural and institutional conditions such as group size, timing of payment (Fischer & Qaim, 2011), group leadership, and cohesiveness of the groups (Paalhaar & Jansen, 2011). Past studies recommended further research, particularly to examine both successful and unsuccessful cases whose lessons can be used to establish strategies for sustainable farmer organizations.

The labour market on the other hand, provides a commercialization opportunity to smallholders. Maertens and Swinnen (2009) argue that smallholders are likely to accrue greater benefits from the commercialization of horticultural farming through the labour market presented by large and medium-scale horticultural farmers in rural areas. Income derived from employment in the high-value supply chains such as export horticulture has been promoted as an engine for economic growth (Swamy, 2004). However, such rural labour markets are limited to specific regions of the country, countering the income distribution effects of rural labour markets. Researchers also criticize the feminization of the labour market for intensifying gender inequality (Barrientos et al., 2003) based on the following reasons. Although women dominate employment in the horticultural industry, they are often discriminated against with respect to wage and the allocation of jobs (Tallontire et al., 2005). They also lack social protection and are frequently discriminated against in various ways including sexual harassment (Hale & Opondo, 2005). Similarly, despite the fact that female farmers and labourers dominate horticultural production they are not equitably rewarded for the labour they invest in the production of agricultural commodities (GoK, 2012).

1.5 Institutional, policy, legal and regulatory frameworks governing the horticultural sub-sector

The legal and regulatory frameworks governing the horticultural sub-sector are composed of acts of parliament and subsidiary legislation (GoK, 2012). Until the recent development of the National Horticultural Policy, the sub-sector was guided by pieces of legislation that focused on agriculture in general (e.g. the Agricultural Act Cap 318 and the Agricultural Produce [Export Act], Cap 319). Unlike the other major agricultural sub-sectors (such as tea and coffee), where both external and domestic trade have been under government control, the marketing of horticultural products has mainly been driven by the private sector. The role of the government has been confined to regulatory services. These services are carried out by two regulatory institutions (KEPHIS and HCDA) that were established through subsidiary legislation (GoK, 2012). Other public institutions that have national mandates on various regulatory aspects include the Kenya Agricultural Research Institute (KARI) and the Pest Control and Product Board (PCPB). In addition, there are private institutions such as the Fresh produce Exporter Association of Kenya (FPEAK) based on voluntary memberships and that emphasise self-regulation and advocacy. The operations of the horticultural industry are also influenced by international legal and regulatory frameworks that provide protocols, standards, conventions, treaties and agreements, and codes of practices for industry associations at the local level (GoK, 2012).

National Horticultural Policy

This policy was established in 2012 with the primary objective of sustaining the growth and development of the horticultural sub-sector in order for it to continue contributing to food security, poverty reduction, employment, and wealth creation. Specific objectives of the policy are: (i) to facilitate increased production of high-quality horticultural produce; (ii) to enhance the provision of the sub-sector's support services (including financing, insurance, research and extension); (iii) to promote value addition and increase domestic, regional, and export trade; (iv) to develop and improve infrastructure to support the horticultural industry, particularly in major production areas; (v) to establish and strengthen institutional, legal and regulatory frameworks in the horticultural industry; (vi) to promote mechanisms for socio-economic and environmental sustainability and other cross-cutting issues; and (vii) to promote investment in horticultural development of ASALS (GoK, 2012). These objectives are aligned with the broader economic goals of the Kenyan Vision 2030 (GoK, 2007a). The drive to establish the policy was primarily fuelled by recognition of the key existing and emerging challenges that hinder the potential of the industry. These include: low market price incentives; multiple taxation systems; the high costs of inputs such as seeds, fertilisers, irrigation water, energy, and air freight; a generally unregulated trading environment that leads to produce "poaching"; and a lack of quality control for local and export produce (GoK, 2012). The policy recognizes the need to develop the domestic market, agro-processing, packaging and quality standards, as well as the potential for the improvement of production and post-harvest practices.

Horticultural Crop Development Authority (HCDA)

The HCDA was established in 1967 through an act of parliament as a government parastatal. Its mandate is to promote the development of horticultural crops, by licensing exporters and disseminating information on horticultural marketing. The authority offers technical and advisory services, including organizing farmer groups and contract farming, supporting horticultural training, providing industry research and information on international trade relations, food safety, environmental protection, and social welfare. The authority provides institutional frameworks to guide the business relationships between exporters and smallholders. One of these frameworks is the "HCDA Code of conduct," which provides guidelines for buyers and the sellers in order to promote good business practices, including the development of legally binding contracts. Another recent framework is the "HCDA order 2011," which was released as Kenya Gazette Supplement No. 166 on December 24, 2011 under "The Agriculture Act Cap 318." The objective of this order is to ensure healthy competition between exporters while maintaining high produce standards for exports. The order requires that all stakeholders involved along the export horticultural supply chains be registered, including the producers, intermediaries, and exporters. Although exporters are in favour of this order, they doubt that the HCDA has adequate capacity to carry out this mandate because the institution has been criticized before for failing to provide useful market information to farmers and commercial buyers (Minot & Ngigi, 2004).

The Kenya Plant Health Inspection Service (KEPHIS)

This institution became operational in 1997 under the Plant Protection Act (Cap 324). The role of KEPHIS in the horticultural sub-sector is to inspect and certify imported and exported horticultural produce and to issue the required sanitary and phytosanitary certificates for export consignments. In addition to the mandatory certification, KEPHIS acts as a third party certification body for voluntary (private) standards such as Maximum Residue Levels (MRS) and GlobalGap.

Fresh produce Exporter Association of Kenya (FPEAK)

FPEAK is a non-governmental organization formed in 1975. The main tasks of this organization include market research (especially of international markets), representing exporters before the government, serving as a liaison to research and regulatory organisations, and supporting small-scale farmer schemes (Minot and

Ngigi, 2004). FPEAK developed KenyaGap, a standard that has been benchmarked in the GlobalGap standard and is recognised by European importers as an equivalent standard, ensuring that Kenyan horticultural products are produced in an environment that protects workers' health and safety without compromising the quality and safety of the products. The aim of developing KenyaGap was to provide a feasible and low-cost version of the GlobalGap standards that could easily be accepted and implemented by all farmers at different scales. This initiative is, however, in a pilot stage but is likely to be adopted by many more producers due to the increase in food quality and safety requirements in both international and domestic markets.

Ministry of Agriculture and related parastatals

The mandate of the ministry is to provide extension services to horticultural producers. The ministry has a fully-fledged Horticultural Division that carries out this role. The MoA in collaboration with other stakeholders in this sub-sector has developed the National Horticultural Policy (GoK, 2012). Closely related to the MoA is the Kenya Agricultural Research Institute (KARI), a parastatal established under the Science and Technology Act (Cap 250) with the national mandate of conducting research activities in agriculture and livestock development. The National Horticultural Research Centre, one of KARI's research centers located at Thika, conducts research on horticultural crops. Another relevant body is the Pest Control Product Board (PCPB), a parastatal established under the Pest Control Product Act (Cap 346). The functions of PCPB are to regulate imports, exports, manufacturing, and distribution of products used for controlling pests, and of the organic functions of plants and animals. The organization plays an important role in ensuring that horticultural produce meets the standard for maximum pesticide residue control of trade for hazardous chemicals in the Kenyan market.

Other national policy frameworks

Other national policy frameworks and strategies that concern this industry include Kenya Vision 2030 and the Agricultural Sector Development Strategy (ASDS, 2010–2020). Kenya Vision 2030 was launched in June 2008 to take over the Economic Recovery strategy (ERS, 2002–2008). It was launched as the new long-term development blueprint for the country, with a vision of transforming Kenya into a globally competitive and prosperous country by 2030 (GoK, 2007b). The aim of Kenya Vision 2030 in the horticultural industry is to enhance existing markets for horticultural products while creating new ones and increasing Kenya's bargaining power in agricultural markets (GoK, 2007a). ASDS was launched in 2010 to accelerate the gains made by the Strategy for Revitalizing Agriculture

(SRA, 2004–2008). In the horticultural industry, ASDS recognizes the challenges and opportunities facing the industry and outlines various interventions to overcome those challenges as well as exploit feasible opportunities (GoK, 2010a).

International regulatory public and private frameworks

The Kenyan export industry depends heavily on European markets. Past food safety scares have led the governments of the EU and retailers to enact strict food safety standards that cover topics such as pesticide residue limits, traceability, farm and packaging hygiene, worker safety and environmental sustainability (Jaffee et al., 2005; Asfaw, 2008). To ensure that the food-safety standards are met, various public and private quality control systems have been developed. Exporters are required to observe most of the public regulatory standards, while the private standards are dependent on the retailers importing the produce. Henson and Reardon (2005) classify these standards into three broad systems: Good Practices (including good agricultural practices), Hazard Analysis and Critical Control Points Systems (HACCP), and international standards set by the International Organization for Standardization (ISO). In the case of EU, the public agricultural product standards are continuously reviewed due to recurring food safety scares. One of the most recent changes in public standards was issued by the Commission Implementing Regulation (EU) directive No. 1235/2012 on imports of certain feed and food of non-animal origin (EU, 2012). The directive came into force on December 19, 2012, following the emergence of new food-safety risks and non-compliance issues with relevant safety requirements. The directive requires increased official controls to be implemented on particular imports originating from certain countries (EU, 2012). The regulations create new costs for exporters from developing countries like Kenya.

The process of setting private standards involves different actors such as NGOs, industrial associations, corporations, and development organizations (Humphrey, 2006). Private retailers have also increasingly been involved in setting requirements for how food is produced and distributed throughout their supply chains. Two prominent private watchdogs of food quality and safety standards for retailers in the EU are the British Retail Consortium (BRC) standard and GlobalGap (formerly the European Retailer Produce Working Group for Good Agricultural practices, EurepGap). The BRC has been operational since 1996. It covers basic food safety and quality requirements including the HACCP system (Fulponi, 2006). GlobalGap standard provides strategies that reflect a harmonization of the existing food safety, quality and environmental guidelines of the major European retailers (GlobalGap, 2008). The guidelines were established in 1997 by the Euro-Retailer Produce working group in response to increasing consumer concerns over food safety and environmental issues (GlobalGap, 2008). The standard has gained importance all over the world, and has been adopted by horticultural growers in most parts of North and South America, Asia, and Africa. GlobalGap is flexible and permits benchmarking of local standards (Asfaw, 2008). Other private standards that are relevant to the Kenyan vegetable export industry include Tesco's Nature's Choice (UK), Marks and Spencer Field to Fork (UK), Ethical Trading Initiative (ETI), Global Food Safety Initiative (IFS), and ISO 900 and 14000 (Asfaw, 2008). Compliance with these standards arises at different levels of the supply chain. Most of them are stringent and primarily adopted by large-scale producers and exporters. The most commonly known private standard for small-scale producers is the GlobalGap standard. Like public standards, compliance with private regulations creates new transaction costs for producers and traders.

1.6 Research methods

1.6.1 Study area

The current study involved a follow-up field visit to a survey that was conducted by the ICIPE during the 2005/2006 cropping season for their project on the Economic Impact Assessment in Horticulture (Asfaw, 2008). The survey was purposefully conducted in five major vegetable producing districts. These districts are located in Central (Nyeri, Murang'a and Kirinyaga) and Eastern (Makueni and Meru Central) provinces of Kenya. These selected districts represent approximately 50% of the smallholders that produce vegetables for the export market (Mithofer et al., 2008) and also have the highest horticultural commercialization incidences for locally consumed vegetables (Sindi 2008). The locations of these districts are shown in Figure 1.5. These districts are endowed with a generally favourable climate for horticultural production, but differ in the intensity and type of vegetable crops produced, agro-ecological characteristics and accessibility (Asfaw, 2008). The region contains different agro-ecological zones that dictate the types of vegetables that can be produced in accordance with the altitude and climate of the area. Meru is located at a higher altitude and mainly produces French beans and snow peas. Nyeri, Kirinyaga and Murang'a are situated at middle altitudes and mainly produce French beans, peas, potatoes, tomatoes, carrots and other domestically consumed vegetables. Makueni is located at a lower altitude and mainly produces Asian vegetables (Asfaw, 2008). In terms of proximity to the capital city Nairobi, Nyeri, Murang'a and Kirinyaga are the closest, and Makueni and Meru are the furthest at a distance of about 250 km from Nairobi. The capital city offers a large market for domestically consumed vegetables owing to its large and rapidly increasing population. Furthermore, most of the vegetables for export are packaged for retail in Nairobi before they are exported by airfreight to their market destinations.

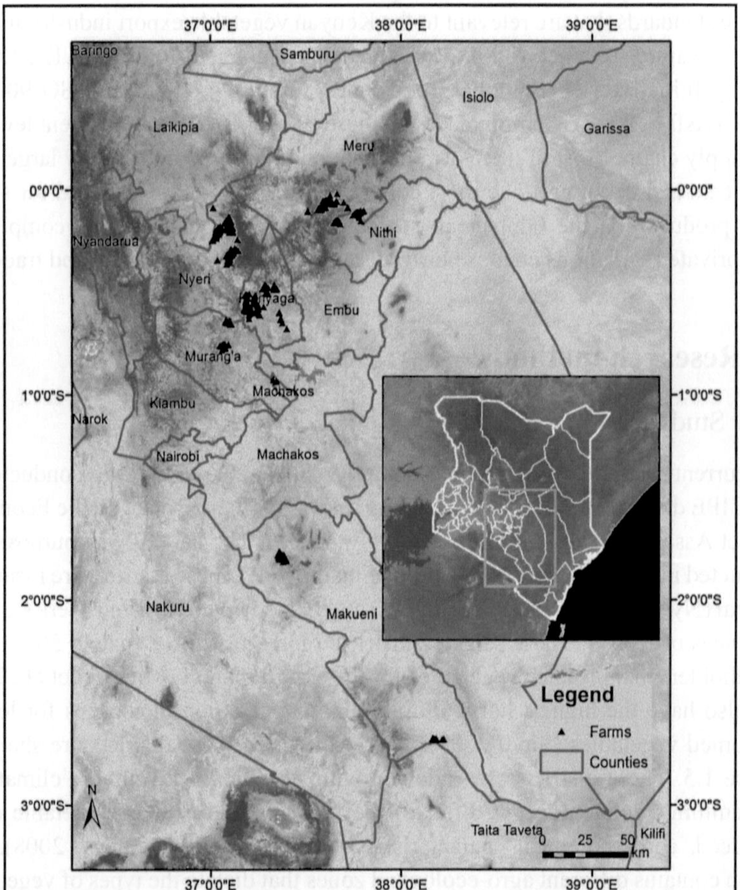

District boundaries: Kenya CBS (2003)[1]
Cartography: Valerie Graw

Fig. 1.5: Location of the study areas in Kenya

---

1 Since 2008, the district boundaries within Kenya have changed due to sub-division of the administrative boundaries. However since this study was a follow-up from efforts that began in 2005, the original names and boundaries of the districts are maintained as they were during the initial phase of the study.

1.6.2 Sampling method

This study utilizes data collected from household interviews conducted in two separate phases that were five years apart. As mentioned in the previous section, the initial survey was carried out by the ICIPE during the 2005/2006 growing season. I conducted the follow-up survey in 2011. A multi-stage sampling procedure was used to select districts, sub-divisions and small-scale vegetable producers. The first stage was the deliberate selection of the five districts mentioned in the previous section. To ensure that all vegetable farmers were included in the sample the Probability Proportional to Size (PPS) sampling method was used. Overall, 21 sub-locations were randomly selected in the five districts using the PPS sampling procedure and a total of 539 vegetable producer households (both export and domestic market vegetable producers) were selected randomly for the interviews. Of these, 439 either specialized in the production of export market vegetables or produced vegetables for both the export and domestic markets, while 100 were purely domestic market vegetable producers (Asfaw, 2008). The list of the 539 households interviewed in 2005/2006 provided a sampling frame for the follow-up survey. Using the PPS sampling procedure, a total sample of 370 households was randomly selected for the second round of interviews.

1.6.3 Data collection

The sample of 370 households was revisited in 2011 by the author, with a sample attrition of 16%. Among the households that were absent from the second interview in 2011, 5.4% of the sample were lost due to death or as a result of moving out of the area, in some cases due to marriage, while the rest were lost due to mismatched household records during the initial survey. Similar to the first round of the survey, data collection during the second round was performed using structured questionnaires administered by trained enumerators, and supervised by the author. Variables of interest included farmer and farm characteristics, including: asset ownership (land, livestock, farm machinery and equipment, consumer durables), buildings, crops enterprises, income and income sources (on-farm and off-farm), access to services (credit, extension services, training), marketing channels, social capital indicators such as producer group membership and GlobalGap compliance. In addition, the geographical information system (GIS) coordinates for each of the interviewed households were utilized to generate satellite-image derived rainfall estimates (RFE 2.0) from the US Agency for International Development (USAID) Famine Early Warning Systems Network (FEWS NET) archives. Several rainfall estimation instruments or techniques that approximate precipitation using cloud cover changes are used to generate 0.1° (approximately, 8km) daily rainfall estimates, which are then aggregated to dekadal (10-day period) rainfall estimates.[2] This method was used to generate rainfall data for each household in the sample for years 2004, 2005, 2009 and 2010. The data was then validated using rainfall data obtained from the Kenya Meteorological Department's weather stations located in or near the study sites. The locations of the farms where the interviews took place are shown Figure 1.5. In addition to the household surveys, focus group discussions were conducted in order to capture the general perceptions of farmers regarding the dynamics of household participation in the commercialization of horticulture during the period between the two survey rounds. A third and final follow up effort was conducted to investigate further the institutional arrangements that promote smallholder market opportunities.

2 The Rainfall Estimates (RFE 2.0) are produced using an interpolation method that combines data from Meteosat and the Global Telecommunication System (GTS), and includes warm cloud information for the dekadal estimates. In addition, the RFE 2.0 uses the Special Sensor Microwave/Imager (SSM/I) on board Defense Meteorological Satellite Program satellites, and the Advanced Microwave Sounding Unit (AMSU) on board NOAA satellites (http://earlywarning.usgs.gov/fews/africa/web/readme.php?symbol=rf, accessed on July 11, 2013). These rainfall estimation data have been used in past studies of climate change concerns (seehttp://earlywarning.usgs.gov/fews/reports.php)

Chapter Two

2. Determinants and Constraints of Smallholder Participation in the Commercialization of Horticulture: An Application of Panel Data to Vegetable Producers in Kenya

2.1 Introduction

Paricipation in commercialization by smallholders can be both a consequence as much as a cause of economic development and poverty reduction, especially in developing countries (Boughton et al., 2007; Barrett, 2008:). Through commercialization, households are provided an opportunity to specialize according to their comparative advantage (Barrett, 2008). The underlying principle of comparative advantage is that markets allow specialization in the production of goods or services in which one is already relatively skilled, or that provide the highest returns on land and labor, or other resources. In the case of households market specialization allows them to use their income to buy goods for household consumption, rather than be constrained by the need to produce all of the goods they need (Pingali, 2007; Barrett, 2008). While the concept of comparative advantage can be efficient in a well-functioning market, commercialization by smallholders in developing countries is typically constrained by risks (production and marketing risks) and high transaction costs that in many cases cause market failure.

In Kenya, the horticultural sub-sector is among the leading contributors to the agricultural GDP. However, the domination of this subsector by smallholder farmers is threatened as a result of emerging challenges that are evolving along with new production and market opportunities and other market forces. In recognition of the need to sustain the industry's growth and development by overcoming the factors hindering it from realizing its potential f, the government of Kenya enacted the National Horticultural Policy (GoK, 2012), with the aim of addressing the input, production and marketing dimensions of the horticultural industry. This study is intended to contribute to the overall objective of the horticultural policy development by identifying constraints and determinants of market participation among smallholder horticultural farmers in rural areas of Kenya.

Previous studies that have attempted to model smallholder decisions about participation in the commercialization of horticulture have used static frameworks that have failed to allow changes in decisions over time that could be influenced by evolving market forces, institutional innovations, or other changes(such as McCulloch & Ota, 2002; Minot & Ngigi, 2004; Omiti et. al. 2007). This shortcoming is overcome in the current study by utilizing data from two waves of selected vegetable producing regions in Kenya in 2005 and 2010 containing information on 309 households. Furthermore, previous studies have not captured weather-related indicators, which are important determinants of the marketing behaviour of farmers. The structure of the current data allows characterization of the trends in output market pathways for the commercialization of smallholder horticultural farming over the 5-year period. In addition, it permits the assessment of factors that determine the choice and extent of commercialization along those market pathways over time. The determinants of different vegetable output market pathways - export and domestic markets – are further distinguished because they have different requirements instead of focusing ona single market as some past studies have done (for example McCulloch & Ota, 2002; Rao & Qaim, 2011). Other studies have analyzed the determinants of simultaneous participation in both markets (for example Olwande & Mathenge, 2011). A two-step estimation similar to the Heckman two-step framework is applied to the panel data to control for individual farm-level heterogeneity within the context of both international (export) and domestic market vegetable producers in selected areas of the eastern and central regions of Kenya. In addition to the characterization of the dynamics of smallholder output market participation, the study applies a multinomial logit model that use the panel data structure to capture unique determinants of different market pathways of participation behaviour. This is done by defining the different market participation choices followed by the farmers during the two survey rounds.

2.2 Theoretical and empirical review of determinants of the commercialization of smallholder horticultural farming

The earliest strand of the literature on the determinants of smallholder market participants used farm household models to explain their market response to changes in relative prices (Strauss, 1984). The studies that followed focused on understanding the role of transaction costs and market failure in smallholder decision making (for example De Janvry et al., 1991; Fafchamps, 1992; Goetz, 1992). These studies developed theoretical household models whose empirical application concluded that transaction costs create barriers for household participation in crop markets and that food and labour market failures are key constraints to participation in the markets. The role of transaction costs has also been theoretically and empirically tested in later studies (Omamo, 1998; Key et al., 2000; Henning & Henningsen, 2007; Barrett, 2008; Voors & D'Haese, 2010).

Other major contributions in the literature of commercialization of agriculture include a review of case studies conducted in ten countries in Africa, Asia, and Latin America - edited by von Braun and Kennedy (1994). In this review, the endogenous and exogenous forces of commercialization are clearly demonstrated. Endogenous drivers are related to farm and farmer characteristics including resource endowment (social, physical, human and financial capital) and household characteristics such as dependency ratio, household size and age and gender of the household head. Other endogenous factors highlighted in past studies include access to information (Omiti et al., 2007), household asset ownership (Heltberg and Tarp 2002; Boughton et al. 2007), financial savings and their substitutes as well as social capital networks, and group membership (Moti et al., 2009; Okello et al., 2009).

Important exogenous factors driving the commercialization of agriculture include rural tourban population shifts, population growth, globalization, technological change, rising per capita income, changes in consumer preferences, increased awareness of nutrition, and changes in macroeconomic or trade policies (von Braun & Kennedy, 1994; Jaffee, 2005; Pingali et al., 2005; Ngugi et al., 2006; Omiti et al., 2007; Sindi, 2008; Virchow, 2008). For example, the globalization of food production has positively influenced the global and interconnected production and distribution of food. In contrast, it has also brought with it concerns of food quality and safety, leading to the development of many public and private food production and marketing practices that may impede the commercialization of agriculture (Jaffee, 2005; Adekunle et al., 2012).

Climate change, as unpredictable annual rainfall patterns and temperature changes, has been cited in recent studies as a factor that increasingly determines the types of crops that farmers choose to commercialize (Virchow, 2008; Kristjanson et al., 2012). Attitude regarding risk has also been documented as a determinant of the extent of agricultural commercialization at the household level (Fafchamps, 1992; von Braun et al. 1994; Dercon, 1996; Ellis 2000). In addition, policies that enhance commercialization have also been recommended in past studies. These include government investment in extension services and research, secure property rights to land and water, improvement of transportation and communication infrastructure, upgrading rural markets, credit services, and other public goods such as better education, health, and sanitation services (von Braun et al., 1994; Pingali & Rosegrant 1995; Minot & Ngigi, 2004; Pingali et al., 2005).

An empirical analysis of the determinants of commercialization by smallholders must address the problem of self-selection. The selection problem arises because households face different commercialization decisions- a discrete decision whether to participate in a market or not and decisions related to the volume or level of produce to sell or to buy, conditional on participation in the markets (Goetz, 1992; Boughton et al., 2007). While all determinants of the volume of the produce to sell or buy affect the discrete decision (whether or not participate in market), the opposite is theoretically not true (Goetz, 1992). Some factors that affect the discrete decision are not expected to affect the continuous decision on the level of market participation (Goetz, 1992). As such, in an empirical analysis variables excluded in the continuous regression model allow identification of market participation equation, which in turn permits accounting for the selection bias (Strauss, 1984).

To address the selection bias, Goetz (1992) applies a two-step model beginning with a probit model on the decision to buy or sell, and then uses a switching regression that allow households to select themselves into buyers and sellers. Alternatively Key et al. (2000) estimate structural supply functions and production thresholds based on censoring models with unobserved censoring thresholds. Their model differentiates between the effects of fixed and proportional transaction costs and finds that fixed transaction costs can be used econometrically to identify the market participation equation. Other studies that have applied Goetz's (1992) approach include Holloway et al. (2001), Heltberg and Tarp (2002), and Boughton et al. (2007). In this study, the discrete participation decision and the continuous commercialization level decision are assumed following Goetz (1992). Further, the approach developed by Key et al. (2000) of differentiating fixed and proportional transaction costs is adopted to identify discrete market participation decision.

This study augments the general literature by adding to the limited studies that use panel data to analyze the determinants of commercialization of smallholder farmers. As the markets evolve, the determinants of market participation by smallholder farmers are equally changing. By using a panel data spanning 5 years and distinguishing participation by different market pathways- *export and domestic markets*- this study identifies factors that are of policy relevance to promoting growth of the horticultural sub-sector in Kenya. Further, by applying a multinomial logit model to analyze the dynamics of the smallholders' market participation decisions, the study highlights the key determinants of smallholder exit from market participation, which have not been addressed in past studies.

2.3 Theoretical and conceptual frameworks

To model the decision to participate in commercialization we begin with the marketed surplus decision analysis framework adopted from Strauss (1984) and Goetz (1992):

$$X_j = X_j(\boldsymbol{p}_j, \boldsymbol{z}, \boldsymbol{k}) \tag{2.1}$$

$$X_j^c = X_j^c[\boldsymbol{p}_j, \eta, A + p_l T(\psi) + \pi(\boldsymbol{p}_j, \boldsymbol{z}, \boldsymbol{k})] \tag{2.2}$$

where X_j refers to household output of good *j*; X_j^c is the household consumption of *j*; ***p*** is a vector of the price for good *j*; *z* is a vector of farm characteristics including fixed inputs, and ***k*** is a vector of production technology parameters. Household characteristics affecting consumption are represented by η, whereas *A* is exogenous income. The price of labour or the wage rate is given as p_l, while *T* is the time available to the household for work and leisure; ψ is a vector of household characteristics determining *T*, and π are profits from output X_j. From Equations (2.1) and (2.2), it follows that the absolute value of marketed surplus (*MSj*) of commodity *j* is given by the difference between production and consumption and is calculated as;

$$|MS_j| = X_j(.) - X_j^c(.) \tag{2.3}$$

Given the absolute value of the marketed surplus $|MS_j|$, Strauss (1984) and Goetz (1992) calculated the elasticity of the marketed surplus of good *j* with respect to its price as follows;

$$\frac{P_j}{|MS_j|}\frac{\partial MS_j}{\partial P_j} = \frac{X_j}{|MS_j|}\frac{P_j}{X_j}\frac{\partial X_j}{\partial P_j} - \frac{X_j^c}{|MS_j|}\frac{P_j}{X_j^c}\frac{\partial X_j^c}{\partial P_j} \tag{2.4}$$

The first term on the right side of the equation is the output elasticity weighted by the ratio of quantities produced to the absolute value of net surplus for production, while the second term is the consumption elasticity value weighted by the ratio of quantity consumed to the net surplus for consumption. Assuming that production and consumption decisions are made simultaneously, the reduced form of marketed surplus is given as;

$$MS_j = MS_j(\boldsymbol{p}_j, \eta, \psi, \boldsymbol{z}, \boldsymbol{k}, A) \equiv MS_j(X_j^{ms}) \tag{2.5}$$

Equation (2.5) shows that the marketed surplus depends on all of the exogenous variables relating to household production and consumption decisions. This implies that the marketed surplus does not only follow the restrictions imposed by the theory of supply and demand functions. Standard supply functions attribute household market response to market prices and ignore the true underlying household-level supply elasticity as characterized by the supply function (Sadoulet & de Janvry, 1995).

Past studies on marketed surplus in rural areas of developing countries reveal that perfect markets rarely exist. Households are located in environments characterized by a number of market failures for some products, for example food (especially perishable foods such as horticultural products or high price risks), and some factors of production such as labour, credit and capital (de Janvry et al., 1992; Sadoulet & de Janvry, 1995; Omamo, 1998; Sadoulet et al., 1998). These market failures create a wide price margin between the lowest price at which a household can sell a good or a factor and the highest price at which it could buy that good or factor. A wide price margin may force a household to choose an 'autarky' position with regards to that good or factor if the subjective price (the price that equates to supply and demand) falls within this range (Sadoulet & de Janvry, 1995). The two fundamental factors that may increase the magnitude of the price margin of horticultural crops in the context of this study are transaction costs and risks related to prices and natural factors such as droughts.

The role of transaction costs in marketing decisions has been discussed in past studies. Transaction costs reduce the price per unit received by households that sell produce in the market and increase the price paid by households that buy the same produce from the market. As such, if some locally consumed vegetable is considered in the presence of transaction costs, the price is given as follows:

$$P_s = P_m + \tau \qquad \text{(buyer)} \qquad \Big| \; \tau = \tau \tag{2.6}$$

$$P_s = P_m - \tau \qquad \text{(seller)} \tag{2.7}$$

$$P_s = P_m \qquad \text{(autarkic)} \tag{2.8}$$

where P_m is the market price for the vegetables and P_s is the shadow price or value of the vegetables to the household. In Equations (2.6) and (2.7) τ represent fixed transaction costs that are equal If τ =0, the household equates its shadow price with the market price. Transaction costs thus create a kinked price schedule reflecting the price margin defined by market prices plus or minus costs to the buyers and sellers respectively as shown in Equations (2.6)–(2.8) (de Janvry et al., 1991). By distinguishing transaction costs into fixed and proportional (variable) transaction costs, Key et al., (2000) provide a solution to identify discrete commercialization participation decisions. The authors demonstrate that market participation decisions are determined by both proportional and fixed transaction costs, while the volume or level of produce marketed (conditional on market participation), are only affected by the proportional transaction costs. Fixed transaction costs can therefore be excluded from the extent or volume of the commercialization decision model, such that the commercialization decision model is identified as follows:

$$Probability\ (commercialization, Ci) = (X_i^{ms}, \tau) \qquad (2.9)$$

Equation (2.9) shows that variables that affect the marketed surplus or extent of commercialization (X_i^{ms}) are also assumed to affect the decision to commercialize or not (*Ci*). However, fixed transaction costs (τ) that affect the commercialization participation decision do not affect the extent of commercialization.

Transaction costs can be defined as the costs of entering into a business contract (*ex ante*) and those of monitoring a contract (*ex post*) (Kirsten et al., 2009). Generally, these transaction costs include information or search costs, negotiation costs, monitoring costs, and certain aspects of transport and storage costs. The transaction costs that are mainly household-related and not commodity specific, for example, those related to market information access and to the ability to negotiate are referred to as fixed transaction costs. Once a household decides to commercialize, it will incur related costs of transferring produce to the market. In horticultural markets, such costs may include certain aspects of transport costs, and barriers such as market fees among other costs. These are referred to as proportional transaction costs. Transaction costs are not always observable. However, available exogenous variables that are likely to determine the outcome of commercialization decision and the extent of commercialization can be used as indicators of transaction costs. In this study, various proxies are used to measure fixed and proportional transaction costs. Fixed transaction costs are given by indicators of access to information, which include access to extension services, ownership of a mobile phone, and access to media through ownership of a TV and/or radio, and to product price information. On the other hand, proportional transaction cost proxies are ownership of a means of transport (car, bicycle or motorcycle), the type of road and distance to the nearest market.

Market participation decisions can also be influenced by risk related factors. Vegetable production is highly dependent on irrigation water in Kenya. Yaro (2013) demonstrates that climate risks shape commercialization decisions of farmers. In a study on the perception and adaptation to climate change by smallholders and commercial farmers in Ghana, the author finds that farmer's knowledge, experiences and perceptions influence their climate adaptation strategies. This includes diversification into different farm enterprises and investing in uncertain physical and economic conditions to reduce potential losses associated with climate risk. Similar observations are highlighted by Hertel and Rosch (2010) in their exclusive literature review on climate change, agriculture and poverty. In the empirical application, we therefore sought to control for climate risk. Since the period between the years of survey are few to warrant climate change measure, we refer it as weather risks. We measure weather risk using a coefficient of variation (CoV) for the temporal variability of rainfall as a measure of relative humidity (precipitation) in different months of the year. The CoV is the percent ratio of standard deviation (*SD*) of rainfall series to the mean (*X*). The CoV is preferred to annual mean rainfall for determining relative wetness or dryness especially when the monthly rainfall for different households or rainfall stations varies (Mishra, 1991). Bronikowski and Webb (1996) also confirm the use of CoV as an appropriate measure for behavioral study of rainfall variability. The CoV measure of weather risk is calculated as follows:

$$CoV_{it} = \left[\frac{SD_{it}}{Mean_{it}}\right] * 100, \qquad t = 1,\ 2,\ i = 1, 2, \ldots .n \text{ (households)}$$

An annual average CoV is calculated based on monthly rainfall estimates either at household level as explained in Section 1.6.3.

2.4 Data and descriptive analysis

The study utilizes two-period panel data collected in five districts purposively selected from the two major vegetable producing provinces (namely Nyeri, Kirinyaga, and Murang'a of the Central Province and Meru and Makueni districts of the Eastern Province) (Asfaw, 2008; Asfaw et al., 2009; Asfaw et al., 2010). The sampling procedure is explained in Chapter 1 of this study. Similarly, to the 2005/2006 survey, the 2011 survey involved recall data that was collected using a structured questionnaire. The recall data provide data on household demographic characteristics, land use, agriculture production, household assets, income and income sources and market access information including type and characteristics of vegetable markets, access to credit and farmer group membership.

Using the information gathered on the type of vegetable crops and market, households are classified according to the market pathway through which they commercialize their vegetables. *Export market pathway famers* produced vegetables for the international market, while *domestic market pathway farmers* produced vegetables primarily consumed in the domestic market, and *non-sellers* are households that did not produce vegetables for sale.[3] In some cases, households produced vegetables

3 International (export) market vegetables include French beans, snow peas, baby corns and Asian vegetables (including cucumbers, okra, aubergines, chilies, karella, valore

for both markets and thus a third category "*both domestic and export markets*" is defined. Based on this classification, Table 2.1 describes the sample considered for analysis in this chapter. It is important to note that, in this study, although vegetables are classified as those for export and those for the domestic market, some crops that were exclusively exported in the past such as French beans are increasingly being consumed domestically, especially in urban areas. However, the share of the domestic market is very small and was not sufficiently documented to warrant distinction between produce sold domestically internationally. The same applies to vegetables such as garden peas and carrots that are mainly produced for the domestic market but a small percentage of fresh-shelled peas and baby carrots are exported.

Table 2.1: Description of the sample size by district

		Households categorized by market pathway and year of survey							
District	*Total households*	Domestic market sellers (%)		Export market sellers (%)		Both domestic & export markets (%)		Non-sellers (%)	
		2005	*2010*	*2005*	*2010*	*2005*	*2010*	*2005*	*2010*
Kirinyaga	88	31	31	28	18	38	24	3	27
Makueni	26	15	58	38	0	46	8	0	35
Meru	78	19	37	38	23	42	17	0	23
Murang'a	27	30	30	4	26	67	22	0	22
Nyeri	90	27	57	11	6	62	32	0	6
Total	***309***	***25***	***42***	***25***	***15***	***49***	***23***	***1***	***20***

2.4.1 Descriptive characterization of market pathways and their participants

An increase in market participation by farmers is an important indicator of positive economic progress (Reardon & Berdegué, 2002). In the study are, there are various types of traders involved in the export and domestic market pathways. Over 80% of farmers producing vegetables for the export market have contract agreements with exporters. Farmers supplying vegetables to the domestic market typically sell their produce to large and small traders without prior agreements on price or quantity. This market arrangement is popularly known as the spot market. The majority of the export market smallholders are organized in production and marketing groups (PMOs); a concept that has been maintained since the 1990's but that strengthened in early 2000 to facilitate compliance with the GlobalGap standards.

and brinjals). Domestic market vegetables include all other types of vegetables that are not produced primarily for the international market (including tomatoes, cabbages, potatoes, peas, kales, onions, capsicum among others).

A significant number of households (about 20%) left the commercial production of vegetables for both market pathways across all districts.[4] Considering the whole sample, households that specialized in the production of vegetables for the domestic market increased by about 17% (from 25% in 2005 to 42% in 2010), while those that specialized in the production of vegetables for the export market decreased by about 10%. The number of households that diversified into both export and domestic markets decreased by about 26%. A similar trend is observed according to production of specific vegetables (Figure 2.1). For example, French beans producers decreased by about 23% (59% in year 2005 to 36% in 2010). A similar scenario is observed for other types of export vegetables. Producers of snow and snap peas decreased from 20% to 13%, while those producing Asian vegetables decreased from 14% to about 5%.

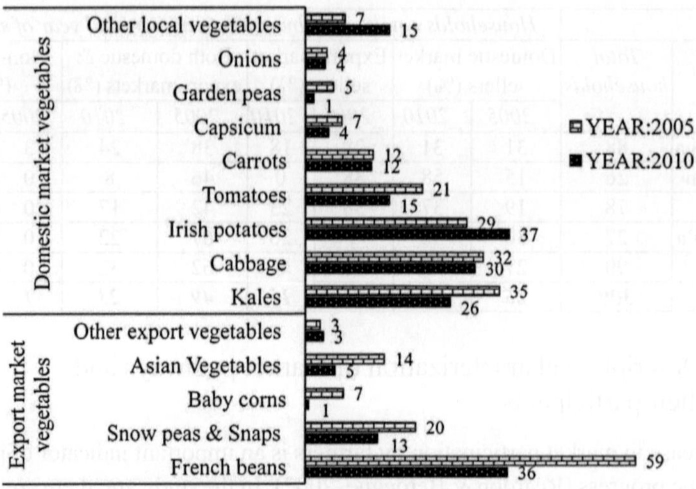

Fig. 2.1: Vegetable crop production among sample households (%)[5]

Domestic vegetable producers exhibit a mixed pattern. Potato producing households increased from 29% in 2005 to 37% in 2010, while cabbage producers

4 Seasonal production is common among smallholder vegetable producers, largely due to climatic conditions. To ensure that this production practice was not confused with farmers who discontinued commercializing their produce, only farmers who o stated they had not produced vegetables for sale for at least one year were considered as having exited the markets.

5 Other local vegetables include; spinach, butternuts, cow peas leaves, African traditional leafy vegetables and pumpkins; while other export vegetables include; beetroots, coriander (dania), leeks, celery, strawberry, and courgettes.

decreased by 2% (from 32% in 2005 to 30% in 2010). A similar pattern is reflected by the number of vegetable types or cultivars produced per household (Figure 2.2). While in 2005 every household produced at least one type of vegetable, about 10% of the households produced none in 2010. About 99% of the households interviewed in 2005 were reported to be selling part of their produce; however, in 2010, about 20% of the same households were not producing vegetables for sale.

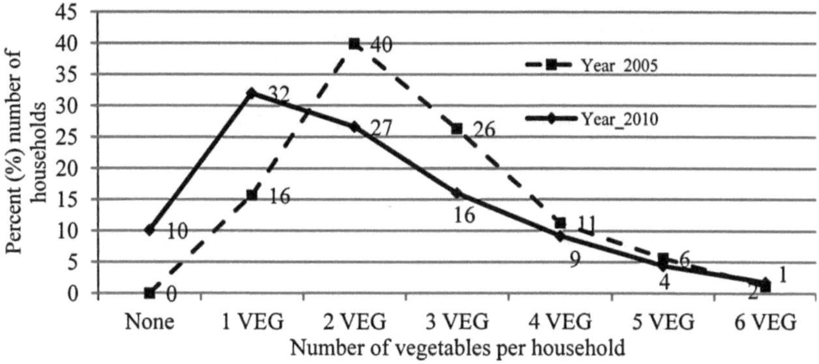

Fig. 2.2: Number of vegetables produced per household among survey respondents

2.4.2 Dynamics of smallholder vegetable marketing pathways

Figure 2.3 shows the marketing pathway dynamics between the two surveys. Out of the 76 households that were specializing in export market vegetables in 2005, only 25% (19 households) specialized in the same crops 2010, 20% (15 households) diversified into domestic market vegetable crops, another 20% shifted to specialization in domestic market vegetables, while 36% (27 households) moved out of the vegetable business entirely. Diversification by export producers in the production of domestic market vegetables could be an indication of spillover-effects of skills gained from producing export vegetables to the production of domestic market vegetables. Since early 2000 most of the smallholder producers for the export market received training on good agricultural practices as groundwork for compliance with the GlobalGap standards (Narrod et al., 2009). However, lack of adequate resources limited compliance with the standards, especially after the withdrawal of initial support, forcing them to stop producing vegetables for the international market (Ouma, 2008). While farmers might have stopped growing vegetables for the export market, the skills gained from the training could have been translated into the production of domestic market vegetables. In addition to the lack of resources for maintaining their position in the export supply chain, the shift of smallholders to the domestic market could be motivated by increased domestic market potential. This is especially so in urban areas due to increasing population and growing demand for vegetables in regional markets (USAID, 2011). This presumption can be supported by other previous studies that have analyzed the prospects for agricultural growth through domestic markets crops such as Diao and Dorosh (2007) and Rao and Qaim (2011).

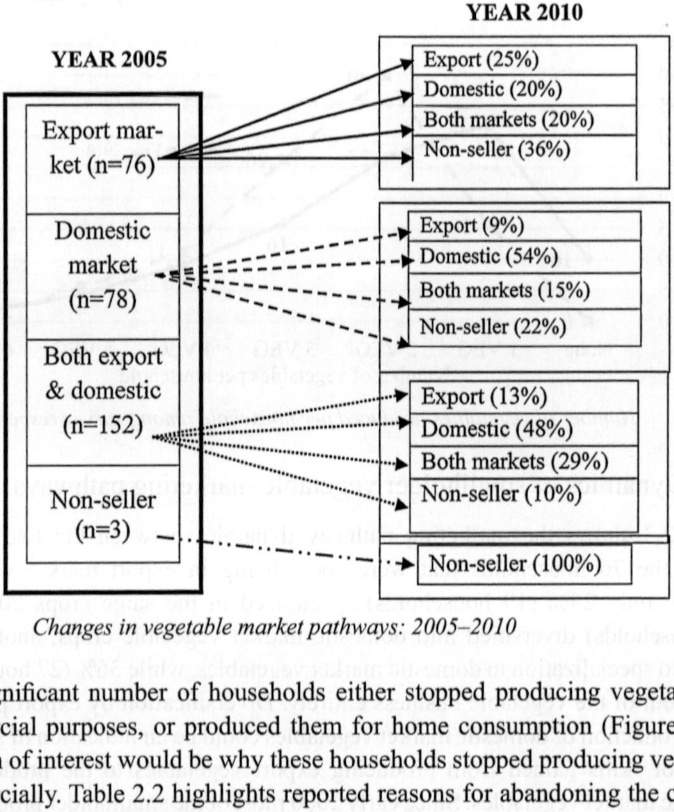

Fig. 2.3: Changes in vegetable market pathways: 2005–2010

A significant number of households either stopped producing vegetables for commercial purposes, or produced them for home consumption (Figure 2.3). A question of interest would be why these households stopped producing vegetables commercially. Table 2.2 highlights reported reasons for abandoning the commercial production of vegetable.

Over 40% of the vegetable non-sellers cited the low returns. This problem was attributed to low farm productivity and to low and highly erratic vegetable market prices. The horticultural markets-both domestic as well as international markets-are subject to high price fluctuations. Farmers with formal contractual arrangements with exporters cited a high incidence (almost 100%) of price defaults. It was also established during the follow-up interviews that most of the exporters are no longer providing binding contract prices but instead prefer flexible contracts that reflect market price fluctuations. This increases market risks to the farmers. In the domestic market, the prices are determined by the market forces mainly because of the perishable nature of horticultural produce, seasonal production, and reliance on rainfall. Seasonality in production and lack of on-farm facilities for vegetable storage results to high post-harvest losses, thus reducing the returns to farmers. High inflation in the recent past could also have contributed to low returns due to the increased costs of inputs such as seeds, fertilizers, and pesticides. These findings are consistent with a recent study on emerging challenges facing the horticultural sub-sector in Kenya (Adekunle et al., 2012).

Table 2.2: Reasons reported for abandoning commercial vegetable production (n=62)

	Percent (%) households
Low returns	43.5
Lack of water for irrigation/low rainfall	35.5
High production costs (inputs/labour costs and other market requirements)	21.0
Lack of ready market (organized market)	17.7
Limited land size	16.1
Lack of interest in horticultural farming	14.5
Labour constraints	12.9
Increased crop pests and diseases	6.5

Farmers also expressed their disappointment with the compliance with the GlobalGap standards, which they perceived as rewarding high prices. They perceived compliance as a "hypocrite business strategy" for the exporters, since they demand compliance with the standards during periods of low demand but purchased produce regardless of compliance during periods of high demand. The cost of maintaining Globalgap standards was also highlighted as adding to the escalation of production costs.

To determine whether the reasons provided for exiting commercial vegetable production of vegetables only affected vegetable enterprises, I examined the livelihood sources of non-sellers during the second survey, which differed across districts according to different agro-ecological conditions. Most of the non-sellers engaged in cereal production in Kirinyaga (63%) and Nyeri (90%). Banana production seems to offer a competitive advantage over vegetable in Meru district, specifically in the lower area of Imenti North. In Meru banana production was adopted by about 33% of non-sellers. Many households that discontinued vegetable production in Murang'a, a district that has the closest proximity to the capital city, Nairobi, shifted to small business operations (50%) and livestock production (40%). Non-sellers from Makueni, which is relatively dry compared to other survey districts, diversified into cereal production (mainly for subsistence), small rural businesses, and manual labourers.

Over 50% of the farmers who maintained their production status until the 2010 survey also reported having simplified their production portfolios compared to 2005. They highlighted similar challenges as those given by households that abandoned production, including shortage of water for irrigation (46%), the high cost of inputs including labour and maintaining GlobalGap standards (26%), increased crop pests and diseases (25%), and diversification, implying less land allocated to vegetable production (17%). Shortages of water for irrigation could reflect negative weather risks that farm households face in many parts of the world. Recent studies on climate change in Africa present mixed conclusions. While some studies predict increased rainfall (Thornton et al., 2008), others show a drying trend or no change rainfall at all (Williams & Funk, 2011). The latter may be true for parts of the study areas, for example in the Makueni district.

The exit of smallholder farmers from the export market is consistent with findings of previous studies in Kenya (McCulloch & Ota, 2002; Jaffee, 2003; Okello, 2005; Ouma, 2008; Ashraf et al., 2009; Okello et al., 2009; Muriithi et al., 2010a). McCulloch and Ota (2002) relate the decline of smallholder participation in the export market partly to the tightening of European regulations on pesticide use and the costs of collecting output from multiple smallholders. Other studies with similar conclusions include Dolan and Humphrey (2000), Jaffee (2003), Okello and Swinton (2007), Nimenya et al. (2009), and Muriithi et al. (2010a). Another reason for the decrease in smallholder participation in the export market is the lack of trust that farmers will make specific investments and that the exporters will purchase produce small-scale farmers (Ashraf et al., 2009). A decrease in smallholder participation in the high-value market has also been observed in other countries (for example Maertens & Swinnen, 2009 in Senegal)

2.4.3 Characteristics of smallholder vegetable growers

Table 2.3 presents a comparison of selected demographic and socio-economic characteristics between export and domestic households. Further, the analysis compares households specializing in the production of either of these markets and those supplying jointly to both markets. Further comparison with the non-seller households is provided in Appendix 2B. On average, the age of the household head for those supplying the export market is lower. The age of the household head is assumed to reflect risk preference and the quality of family labour. Younger farmers are expected to be more likely to adopt risky and labour intensive farm enterprises that have higher expected income such as high-value export vegetables

such as export vegetables. From a different perspective, older age may imply more experience and hence better management practices that could favour the production of high-value crops. The former, however, applies to this study as shown in Table 2.3. Ownership of agricultural implements and livestock are not significantly different across different groups of vegetable producers.[6] A significantly larger proportion of export market suppliers own fertile land (40%), while more domestic market suppliers are engaged in rural medium and small businesses.

The average distance to the nearest market town is significantly lower for export market suppliers than domestic market suppliers. It was observed during the fieldwork that export vegetables are harvested and thus collected more frequently (on average twice per week) in comparison to domestic market vegetables (mostly collected once or on longer intervals). Export market suppliers are therefore likely to be found nearer the market town with relatively better road infrastructure. As expected, most export market suppliers are members of PMOs (63%) and complied with the GlobalGap standards (56%). A significant proportion of export vegetable producers received credit.

Corresponding to the earlier observations that farmers who dropped out of commercial vegetable production are often engaged in cereal production, non-sellers have a significantly larger proportion of their land allocated to maize production compared to those specializing in export market crops and those supplying to both markets (see Appendix 2B). Although not significant, non-sellers have larger shares of total land owned and cultivated compared to households that are involved in the commercial production of vegetables (see Appendix 2B). This may encourage non-vegetable producing households to cultivate maize, which is not as land intensive as vegetables (lower output per unit area). Maize prices in Kenya have remained high over the past few years, mainly attributed to the high variability of rainfall. Furthermore, increases in the price for imported maize as a result of the global food price crises of 2008 and 2011 could also account for high maize prices in the domestic market (Dillon & Barrett, 2013). Consequently, farmers would be compelled to produce more maize, not only to tap the benefits of high market prices but also for household use to increase food security.

---

6 The agricultural implements and livestock asset indices were constructed using the Principal Component Analysis (PCA) approach following Irungu (2002), Henry et al. (2003), Rutstein & Johnson (2004), and Zeller et al. (2006). These are two of the five key categories of non-land assets identified as important for the study area. Livestock assets include all types of livestock (cattle, small ruminants, draft animals and poultry), while Agricultural assets include hosepipes, water pumps, sprinklers and insecticides' pumps). Other categories are explained in the subsequent Chapter.

Table 2.3: Characteristics of households by vegetable output market

	Total observations (n=618)		Export market suppliers (E) (n=122)		Domestic market suppliers (D) (n=208)		*t-test (E-D) diff.*	Both markets suppliers (B) (n=223)		*t-test (E-B) diff.*	*t-test (D-B) diff.*
	Mean	SD	Mean	SD	Mean	SD		Mean	SD		
Household head gender (*1=male*)	0.91	0.29	0.95	0.22	0.90	0.30	*0.05*	0.91	0.29	*0.04*	*–0.01*
Household head age (years)	48.7	12.5	45.0	11.2	50.7	12.5	*–5.7****	46.9	12.0	*–1.85*	*3.87****
Head education (years of schooling)	8.69	3.46	8.70	2.88	8.62	3.95	*0.08*	8.87	3.20	*–0.17*	*–0.25*
Household size (*adult equivalent*)	4.91	2.23	4.74	2.14	5.09	2.32	*–0.35*	4.87	2.19	*–0.13*	*0.22*
Oxen ownership	0.03	0.18	0.06	0.24	0.02	0.13	*0.04***	0.03	0.16	*0.04**	*–0.01*
Agricultural implements (*index*)	1.04	0.42	1.03	0.44	1.07	0.41	*–0.04*	1.07	0.40	*–0.04*	*–0.01*
Livestock (*index*)	0.13	0.14	0.13	0.16	0.14	0.16	*0.00*	0.13	0.12	*0.00*	*0.00*
Fertile land (*dummy*)	0.28	0.45	0.40	0.49	0.31	0.46	*0.09**	0.20	0.40	*0.20****	*0.11****
Off–farm employment (*dummy*)	0.31	0.46	0.27	0.45	0.33	0.47	*–0.06*	0.34	0.48	*–0.07*	*–0.01*
Remittances (*dummy*)	0.27	0.45	0.24	0.43	0.32	0.47	*–0.08*	0.20	0.40	*0.04*	*0.12****
Business (*dummy*)	0.29	0.45	0.24	0.43	0.34	0.47	*–0.10**	0.24	0.43	*0.00*	*0.09***
Distance to nearest market town (*Km*)	3.96	4.57	2.84	3.27	3.93	4.10	*–1.10***	5.18	5.61	*–2.34****	*–1.25****
Road type (*1=good*)	0.50	0.50	0.48	0.50	0.55	0.50	*–0.07*	0.43	0.50	*0.04*	*0.11***
Total land owned (*acres*)	3.14	4.45	2.92	5.46	3.49	5.26	*–0.57*	2.79	2.71	*0.13*	*0.70**
Land cultivated (*acres*)	2.22	2.10	2.11	1.88	2.21	2.61	*–0.10*	2.19	1.68	*–0.08*	*0.02*
Access to credit (*dummy*)	0.38	0.49	0.57	0.50	0.31	0.46	*0.26****	0.35	0.48	*0.21****	*–0.05*
Annual rainfall (*mm, lagged*)	1073.7	116.7	1084.1	123.2	1059.0	111.8		1109.7	89.4		
Covariance of variation (*monthly precipitation, %*)	54.2	8.0	57.3	9.3	53.6	7.38		52.1	7.16		

	Total observations (n=618)		Export market suppliers (E) (n=122)		Domestic market suppliers (D) (n=208)		*t-test (E-D) diff.*	Both markets suppliers (B) (n=223)		*t-test (E-B) diff.*	*t-test (D-B) diff.*
	Mean	SD	Mean	SD	Mean	SD		Mean	SD		
Vegetable contact (*dummy*)	0.46	0.50	0.82	0.39				0.76	0.43	*0.06*	
Member farmer group (*dummy*)	0.43	0.50	0.63	0.48				0.68	0.47	*–0.05*	
Member farmer group (*years*)	1.80	3.76	3.43	5.22				2.14	3.39	*1.28****	
GlobalGap compliant (*dummy*)	0.30	0.46	0.56	0.50				0.49	0.50	*0.07*	
Extension contact (*dummy*)	0.53	0.50	0.73	0.45	0.55	0.50	*0.18****	0.69	0.46	*0.04*	*–0.14****
Transport facility (*dummy*)	0.61	0.49	0.59	0.49	0.55	0.50	*0.04*	0.48	0.50	*0.11***	*0.08**
Own TV (*dummy*)	0.49	0.50	0.43	0.50	0.53	0.50	*–0.11**	0.47	0.50	*–0.04*	*0.06*
Own Radio (*dummy*)	0.94	0.24	0.93	0.26	0.97	0.18	*–0.04**	0.93	0.25	*–0.01*	*0.03*
Own phone (*dummy*)	0.78	0.41	0.70	0.46	0.84	0.37	*–0.13***	0.74	0.44	*–0.04*	*0.09***
French beans price ($/Kg)	0.49	0.16	0.50	0.16	0.47	0.15	*0.02*	0.52	0.17	*–0.03*	*–0.05*
Snow peas price ($/Kg)	0.74	0.11	0.74	0.13	0.74	0.10	*0.00*	0.76	0.11	*–0.02*	*–0.02*
Potato price ($/Kg)	0.19	0.07	0.21	0.07	0.17	0.07	*0.04*	0.21	0.08	*0.00*	*–0.04*
Cabbage price ($/Kg)	0.18	0.18	0.13	0.07	0.22	0.22	*–0.10*	0.18	0.17	*–0.05*	*0.05*
Maize price ($/Kg)	0.26	0.07	0.24	0.09	0.26	0.01	*0.02**	0.24	0.01	*–0.01*	*0.01*
Maize area (%)	21.86	18.67	19.50	19.49	24.46	18.68	*–4.95***	19.09	17.80	*0.42*	*5.37****
Potato area (%)	8.23	15.48	1.81	5.85	11.87	19.31	*–10****	9.78	14.92	*–7.97****	*2.09*

Note: Significance at 0.01(***), 0.05(**), 0.1(*); the figures are means for continuous variables and proportions for categorical variables. Vegetable contract, membership in PMO and GlobalGap compliant are observed only among suppliers to export and both markets. As such, there are no figures for these variables for the domestic market participants. The sum of export, domestic, and both markets do not add up to the total (n=618). The total figure includes non –sellers (included in Appendix 2B). The exchange rate for 2005 was approximately 75 Kenya shillings (Ksh) per U S $, 79 Ksh/US$ in 2010 and 79.7Ksh/US$ in February 2009 (Base month).

2.4.4 Measuring the extent of commercialization along different market pathways

The extent of commercialization can be examined in simple terms as the cash derived from the sale of vegetables. Commercialization can also be defined using different indices (horticultural commercialization indices, HCI). These are formulated and utilized in this study. First, commercialization as the annual proportion of income derived from vegetable sales out of total household income (HCI_1). Second, commercialization measured as the annual proportion of income derived from vegetable sales out of the total crop sales (HCI_2). Lastly, commercialization as the annual proportion of income derived from the sale of vegetable crops out of total value of vegetable crops produced in the household (HCI_3) (Govereh et al.,1999;Strasberg et al., 1999; Sindi, 2008). The three horticultural commercialization indices (HCIs) can be expressed as follows;

$$HCI\_1_{i,t} = \frac{Gross\ value\ of\ vegetable\ sales_{i,t}}{Total\ household\ income_{i,t}} x\,100$$

$$HCI\_2_{i,t} = \frac{Gross\ value\ of\ vegetable\ sales_{i,t}}{Gross\ value\ of\ all\ crop\ sales_{i,t}} x\,100$$

$$HCI\_3_{i,t} = \frac{Gross\ value\ of\ vegetable\ sales_{i,t}}{Gross\ value\ of\ all\ vegetable\ crops\ produced_{i,t}} x\,100 \quad (2.10)$$

where $i=1, 2, 3\ldots$ n (households) and t=1,…, T (years). The indices measure the degree of market orientation of a household's vegetable production. A high index value implies a higher level of commercialization. Figure 2.4 shows the commercialization trends using the simple definition as annual vegetable sales by output market pathway.

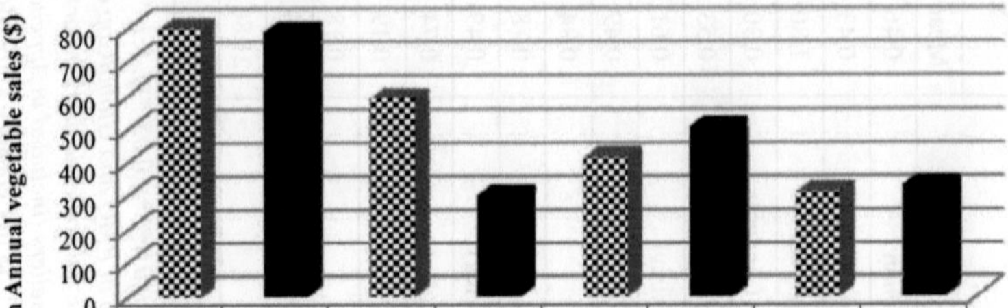

	Export vegetable sales				Domestic vegetable sales			
	Suppliers		Total sample		Suppliers		Total sample	
	_2005 n=228	_2010 n=117	_2005 N=309	_2010 N=309	_2005 n=230	_2010 n=201	_2005 N=309	_2010 N=309
■ Mean	796	786	587	298	412	505	307	328

Fig. 2.4: Mean annual income from vegetable sales through different market pathways

Considering the sub-sample of households that supplied some vegetables for the export market in each year, vegetable sales decreased insignificantly (from US$796 in 2005 to US$786 in 2010), but a much higher decrease of about US$289 (from US$587 in 2005 to US$298 in 2010) occurred for the total sample (n=309).[7] On the other hand, average earnings from domestic market vegetables increased by about 23% (from US$412 in 2005 to US$505 in 2010) among households who supplied those vegetables. The results presented in Figure 2.4 confirm that, on average, vegetables supplied to the export market earned higher income than those supplied in the domestic market.

Figure 2.5 presents the trend of returns from the export market versus that of the domestic market pathway, using different measures of commercialization (HCIs). Among the households that were exclusively supplying vegetables to the export market in 2005, vegetables contributed about 39% of total household income (HCI_1), while in 2010, the share dropped to 26%. The contribution of domestic market vegetable sales to household income increased from 18% in 2005 to 21% in 2010. The contribution of vegetable sales to total household income, among households supplying vegetables jointly to the export and domestic markets, decreased by about 3%. This finding concurs with earlier studies that report declining returns from vegetables as well as other agricultural marketed products (Olwande & Mathenge, 2011).

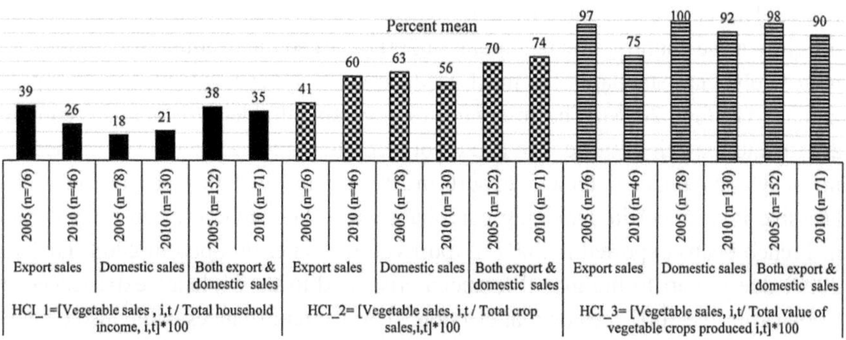

Fig. 2.5: Horticultural commercialization index by market pathway

7 Values for the 2010 survey were deflated while those of 2005 are inflated using consumer price index (CPI) to a common year (February 2009) using the consumer price index data available from the Kenya Bureau of Statistics (http://www.knbs.or.ke/consumerpriceindex.php). The Exchange rate for February 2009 was approximately 79.70 Kenya shillings (Ksh)/US$. This is the rate used in the above and subsequent analysis.

On the contrary, export market vegetable sales remained important among the other crop sales as shown by the increase in the proportion of vegetable sales of total crop sales (HCI_2) from 41% to 60%. This may imply that households that continued to produce export market vegetables are more dependent on vegetables than other crops. This may be true in Kirinyaga district where farmers have limited cash crop options due to small parcel sizes compared to, for example, Meru where tea and coffee are produced alongside commercial vegetables. This result should not be interpreted as an increase in the production of export vegetables, however, as households could have reduced their commercial crop portfolios and engaged in other livelihoods. On the other hand, the contribution of domestic market vegetables out of total crop sales decreased from 63% to 56%. Overall, vegetable sales contributed an average of 60% (not shown in this graph) of total crop sales and about 30% of total household annual income, which demonstrate their importance to livelihoods of smallholders. For HCI_3 the contribution of export vegetable sales among households specializing in the production of these vegetables decreased by about 22%. The proportion of domestic market vegetables decreased by about 8% among exclusive suppliers to this market and both markets suppliers.

Generally, the commercialization of export vegetables among smallholders decreased whereas domestic market sales increased, albeit by a small percentage. However, it is important to note here that the 5-year period between 2005 and 2010 was important relative to the performance of the export horticultural sector in Kenya. This can be illustrated using the case of French beans, a major export vegetable crop in the country. The beginning of 2005 marked the deadline for compliance with GlobalGap standards for most producers supplying vegetables to European retailers (Asfaw, 2008). The re-organization of smallholders to facilitate compliance led to the exit of those who could not meet these requirements. The immediate impact was a reduced volume of exports in 2005 through to 2006 (see Appendix 2A). Massive investment in achieving compliance with GlobalGap raised the volumes of French beans exported in 2007. Export volumes later dropped in 2008 (HCDA, 2009) due in part to the global financial crisis and to the violence experienced in Kenya after the 2007 presidential election.[8] These circumstances could also explain the shift of smallholder farmers supplying export market vegetables to domestic market vegetables and other farm products with more reliable markets.

---

8 The negative effect by the 2007 post-election violence is supported by a recently published empirical study by Muhammad et al. (2013) who find a significant negative impact of the violence and instability on the EU flower imports from Kenya.

2.5 Empirical models

Three models are developed for the choice of market pathways and the extent of participation. The first and second are two-step models as explained in section 2.3, that address endogeneity due to self-selection and omitted variables that have largely been ignored in past studies. In addition, a multinomial logit model is used to explore the dynamics of smallholders' participation among different market pathways. By applying the multinomial logit model, we separate individual households into five categories generated from the description of households given in Figure 2.3.

2.5.1 Estimating the choice of market pathway and the extent of commercialization

As described in section 2.3, participation in the commercialization of horticulture involves a two-step decision problem: the household decides whether or not to commercialize and then establishes the extent or level of commercialization conditional on participation. Because vegetable sales through either domestic or export markets are only observed for a subset of the sample, sample selection problem exists. The problem can be regarded as an incidental one since the level or extent of commercialization can only be measured if a farmer participates in the commercialization of horticulture, and households with sales observations are not likely to be a random sub-sample. In the context of panel data, Heckman's (1976) two-step regression framework is extended to address endogeneity due to household self-selection behavior as non-sellers, domestic market producers, export market producers, and producers for both markets. Each of the two principle market pathways (export and domestic markets) has a participation equation and a regression equation. Since the two markets are not mutually exclusive such that there are some farmers who participated in both markets, a third estimation is made for such households.

To illustrate the estimation procedure, let the regression equation for the level of commercialization be as follows;

$$Y_{ijt} = \boldsymbol{X}_{\mathrm{ijt}}\beta_I + C_i + \mu_{ijt} \quad i = 1, 2\ldots..n,\ t = 1,2\ldots..T,\ j = market\ pathway \quad (2.11)$$

where Y_{ijt} is the level of the commercialization measure for household *i*. The market pathway is represented by *j* (*j*=export market, domestic market, both markets), and *t* is the time period. ***X*** represents a set of observed independent variables that influence the level of commercialization, C_i represent unobserved household time invariant variables (unobserved effects) and μ is the observed error. Y_{ijt} is observed only if the binary selection indicator S_{ijt} is positive, such that if S_{ijt}=1 we observe vegetable sales for market pathway *j*. Suppose that for each *t*, S_{ijt} is determined by a probit or logit equation:

$$S_{ijt} = 1(\boldsymbol{W}_{ijt}\gamma_{it} + \upsilon_{ijt} \geq 0) \qquad \upsilon_{ijt} \mid W \sim Normal(0,1) \tag{2.12}$$

where $\boldsymbol{W}$ is a vector of the observed variables that influence S_{ijt}. The error term υ_{ijt} is assumed to be independent of W_{ijt} (and therefore X_{ijt}), and have a random normal distribution. However μ_{ijt} and υ_{ijt} may be correlated providing biased estimates if Ordinary Least Squares (OLS) is used to estimate Equation (2.11).

$$Corr(\mu_{ijt}, \upsilon_{ijt}) = \rho \neq 0 \tag{2.13}$$

In the sample, farmers switched from one pathway to another in the sense that farmers did not stay in one market pathway over the 5 year period (see Figure 2.3). Some dropped out permanently, whereas others changed to a different market pathway. As such, the Heckman framework following the Chamberlain approach to panel data models can be used (Wooldridge, 2002, pp. 582–583; Greene, 2012, p. 926). The approach involves two stages; the first stage involves estimating a probit model of S_{ijt} for each t and then saving the inverse mills ratio (non-selection bias), $\hat{\lambda}_{ii?}$, for each observation, such that:

$$\hat{\lambda}_{ijt} = \frac{\varphi(S_{ijt}\hat{W})}{\Phi(S_{ijt}\hat{W})}, \tag{2.14}$$

where $\varnothing$, is the standard normal probability density function and Φ is the standard normal cumulative density function. In the second step, a pooled OLS is applied on the selected sample by adding $\hat{\lambda}_{ii}$ to the regression Equation (2.11) and interaction of itself with the time dummy (T) in the regression equation such that:

$$Y_{ijt} = \boldsymbol{X}_{\mathbf{ijt}}\beta_{I} + \psi_1\hat{\lambda}_{ijt} + \psi_2(d2010_i * \hat{\lambda}_{ijt}) + \mu_{ijt} \tag{2.15}$$

An alternative way to estimate Equation (2.15) is one suggested by Wooldridge (2005). Individual level fixed effect estimation in the second step may be used without interactions between the inverse mills ratio and the time dummy such that Equation (2.15) is estimated as follows:

$$Y_{ijt} = \boldsymbol{X}_{\mathbf{ijt}}\beta + \psi_1\hat{\lambda}_{ijt} + C_i + \psi_2 d2010_i + \mu_{ijt} \tag{2.16}$$

Both estimation procedures are adopted in this study. An important requirement is that $\boldsymbol{X}$ is a strict subset of $\boldsymbol{W}$ such that any element of $\boldsymbol{X}$ is an element of $\boldsymbol{W}$, however, there are some elements of $\boldsymbol{W}$, that are not elements of $\boldsymbol{X}$. This is known as the exclusion restriction (Wooldridge, 2002) or the identification condition (Maddalla 1983). As explained in Section 2.3, following Key et al., (2000) the selection equation is identified by excluding fixed transaction cost indicators/variables from the pooled OLS regression equation. The fixed transaction costs proxies utilized in the analysis includes access to extension services, ownership of a mobile phone, and access to media through ownership of a TV and radio, and district market prices for the most important vegetables. As explained in sections 2.2 and 2.3, ***X*** contains various exogenous variables that are related to the commercialization of horticulture. These include demographic characteristics (such as gender, age and education of the household head and household size), farm characteristics (size of land owned, and cultivated, and land quality), asset ownership (agricultural assets and livestock), income sources (off-farm, remittances, business) and proxies for proportional fixed transaction costs (ownership of a means of transport, distance to the nearest market and condition of the road to nearest market). In addition, proxies for weather shock and weather risk explained in Section 2.3 are included. Time dummies are included in the second stage to capture time varying temporal effects such as technological changes, skill development and other farm and market influences.

2.5.2 Estimating the dynamics of vegetable market participation decisions

An issue that arises when studying market participation by smallholders over time is that there are more market pathways that they might follow. While some households maintain one market (export or domestic), others shift from one market to another, and some drop out (Figure 2.3). Given this description in Figure 2.3, individual households can be separated into a minimum of five distinct market pathways: export market suppliers, domestic market suppliers, suppliers of both markets, switching suppliers (from either of the three named to another) and 'exitors' (those who drop out of commercialization completely). The first three categories imply households that maintain the same market pathway during the analysis period. Identifying the behaviour and key differences among households in each category provides more information about the dynamics of the market participation decisions than a simple binary analysis of an adoption model. A multinomial logit model using panel data can capture the determinants of each of these distinct household categories. Unlike cross-sectional data analysis, this estimation is able to compare the determinants associated with households that exit from market participation or that switch to another market against those who maintain one market supply pathway. The dynamics of smallholder market participation is overlooked in the estimation described in the previous section.

Participation in either of the market pathways is not ordered. In the case of unordered choices as dependent variables, the selection of either of the choices can be motivated by a random utility model following Greene (2012). The underlying assumption of this utility model is that each household makes its market participation choice for each period according to a latent utility function s^* defined as follows:

$$s^*_{ikt} = \omega_{it}' \beta + \mu_{ikt} \quad (2.17)$$

where S^*_{ikt} is the utility of household i values market participation choice k (k=0,1,….K) for time t, K+1 is the number of market participation choices, $\boldsymbol{w}_{it}$ are observed explanatory variables, $\boldsymbol{\beta}$ is a vector of the parameters to be estimated, and the μ_{ikt} is the error term that may contain both individual household specific and idiosyncratic elements that are not included in $\boldsymbol{w}_{it}$. Consider the multiple market participation pathways (i.e. k=0, 1,…., K choices) and time t=0,1 (t=0 if 2005 and t=1 if 2010), such that k categories can be defined as follows:

$$\begin{array}{llll}
k=1 & if\ s^*_{i1t} > 0 & for\ t=0\ and\ t=1 & \text{Export market suppliers} \\
k=2 & if\ s^*_{i2t} > 0 & for\ t=0\ and\ t=1 & \text{Domestic market suppliers} \\
k=3 & if\ s^*_{i3t} > 0 & for\ t=0\ and\ t=1 & \text{Joint export and domestic market suppliers} \\
k=4 & if\ s^*_{ilt} > 0 & for\ t=0 & \text{Switched market suppliers} \\
 & but\ s^*_{imt} > 0 & for\ t=1 & (l,m=1,2,3\ l \neq m) \\
k=0 & if\ s^*_{ilt} > 0 & for\ t=0 & \text{Exitors } (l=1,2,3) \\
 & but\ s^*_{ilt} \leq 0 & for\ t=1 &
\end{array}$$

The above formulation is dynamic. If characteristics that determine in which category a household falls can be adequately described in the first period, then the estimation can be reduced to single period estimation. Following the random utility model assumption that the *ith* farmer is faced with K market participation choices, the household would select the option k that maximizes its utility. Given K choices and specific household explanatory variables (w_{it}), the probability for K+1 are determined by the following equation;

$$\Pr(s_i = k \mid w_i) = P_{ik} = \frac{\exp(w_i' \beta_k)}{1 + \sum_{l=0}^{K} \exp(w_i' \beta_k)} \quad k = 0,1,\ldots.K \quad (2.18)$$

where *Pr(.)* is the ith individual probability of choosing the kth pathway conditional on $\boldsymbol{w}$. With K choices, K log-odds are computed. Because the response probabilities must sum to 1, a benchmark outcome (k=0) can be assigned to identify the coefficients in the estimation of different market participation options relative to the benchmark outcome, such that:

$$\Pr(s_i = 0 \mid w_i) = P_{ik} = \frac{1}{1 + \sum_{k=0}^{K} \exp(w_i' \beta_k)} \quad (2.19)$$

The above specification of the multinomial logit model can partially address endogeneity that arises from self-selection of households to different market pathways, since it utilizes characteristics of households from the baseline to describe the current period (Barham et al., 2004). However the endogeneity solution is addressed only for the households that changed their market participation status (*k=0 & 4*), but not the ones that maintained their market status (*k=1, 2 & 3*). This is plausible since the drivers of dynamics of smallholder participation can be identified, especially for the exitors from commercial vegetable production, which has not been addressed in previous studies. Furthermore, the factors that motivate the choices for the status (*k=1, 2 & 3*) are addressed in the previous estimation.

2.6 Econometric results

The regression results for commercialization through the export market and domestic market pathways are shown in Table 2.4 and Table 2.5, respectively. Table 2.6 presents the results for households that supplied both markets. In each table, pooled OLS (Equation (2.15) estimates and fixed effects (Equation 2.16) estimates are provided that account for endogeneity due to self-selection for the extent of commercialization are provided. In addition, the first stage probit model estimates (selection Equation (2.12) are provided by survey year, as estimated to calculate the selection bias term (inverse mills ratio). It is important to note that the estimations of the selection equations (whose results are presented in tables 2.4 and 2.5) contain the whole sample including households that produced for both markets. Although the extent of commercialization is defined in different measures as outlined in section 2.4.3 (Figures 2.4 and Figure 2.5), the HCI_1 (the share of income derived from vegetable sales out of total household income) is the dependent variable. We utilize this index because, unlike the alternative HCIs mentioned in the previous section, HCI_1 reflects the importance of vegetable enterprises to the overall livelihood of a household. The other two indices are measured based on crop and vegetable production, thus excluding other livelihood sources. The index is natural log transformed to standardize the distribution.

2.6.1 Commercialization through the export market pathway

Table 2.4 shows the estimates of the extent of commercialization and the selection equation of the export market pathway. The dependent variable is the logarithmic value of the proportion of export vegetable sales to total annual household income (HCI_1) in columns (1) and (2) and whether a household commercializes or not in 2005 or 2010 in columns (3) and (4), respectively. The null hypothesis that all parameters of interest are simultaneously equal to zero is rejected at the 1% significance level ($p<0.01$) for both the OLS and fixed effects models. The selection models for each year of the survey are also highly significant ($p<0.01$) based on the log-likelihood ratio chi-square test.[9] The R squared for the fixed effects model and pooled OLS regression are 52% and 23% respectively. The estimate for the inverse Mills ratio is statistically significant only in the specification including fixed effects. This provides evidence for self-selection into commercialization being an issue, at least when we control for unobserved heterogeneity across households.

The proportional transaction cost indicators (distance to the nearest market town, road type, and ownership of a means of transport do not appear to have any significant effects on the extent of commercialization through the export market (Table 2.4). In contrast to expectations, the distance to the nearest market town is positive and significantly ($p<0.1$) related to the decision to commercialize through this market pathway in 2010. A possible explanation, based on field observations, is that households near market towns tend to engage in other non-farm activities such as small business unlike those that are farther from market center and whose livelihood opportunities are limited to farm enterprises. Furthermore, because it is costly for farmers to transport harvest to the market, vegetable traders collect produce directly from farms or at designated collection centers nearby. Similar observations are made by Minot and Ngigi (2004). Sindi (2008) also finds a positive association between the distance to the nearest transitable road (a correlate of proximity to the nearest market town) and horticultural commercialization. Sindi explains that rural areas with good conditions for growing horticultural crops have poor road access, especially due to heavy rainfall experienced in those areas. Considering the distance between farm households and the point of sale, Omiti et al. (2007) find an inverse relationship between distance and the percentage of horticultural output sold.

From the selection model, none of these fixed transaction cost proxies are significant except product prices. The prices of French beans and snow peas in 2010 are positively related to the decision to commercialize through the export market. Information related proxies—ownership of a TV, radio, phone and access to extension services—do not have significant effects.

9 Ramsey's test rejects the null hypothesis that there are omitted variables. In the selection model, the link test model rejects the null hypothesis that there are omitted variables. Hosmer–Lemeshow statistic also applied after the selection model to evaluate the goodness of fit of the model.

Table 2.4: Determinants of vegetable commercialization through the export market pathway

	In (Horticultural commercialization index, HCI_1)				*Selection equation (export=1)*			
	Household fixed effects (1)		Pooled OLS (2)		Year_2005 (3)		Year_2010 (4)	
Head gender (1=male)	0.58	(0.54)	–0.17	(0.22)	–0.24	(0.32)	0.57*	(0.34)
Head age (years)	0.001	(0.03)	–0.01**	(0.01)	–0.03***	(0.01)	–0.02**	(0.01)
Head education (years of schooling)	–0.10*	(0.05)	0.001	(0.02)	–0.02	(0.03)	–0.02	(0.03)
Household size (adult equivalent)	–0.35**	(0.17)	–0.07	(0.10)	0.05	(0.15)	–0.17	(0.17)
Household size (squared)	0.02*	(0.01)	0.004	(0.01)	–0.003	(0.01)	0.02	(0.01)
Oxen ownership	–0.81	(0.60)	0.48	(0.38)	0.13	(0.73)	–0.05	(0.58)
Agricultural implements (index)	0.43	(0.36)	0.13	(0.17)	0.17	(0.27)	0.44	(0.27)
Livestock (index)	–1.50*	(0.82)	–2.13***	(0.56)	0.46	(0.90)	1.48	(0.93)
Fertile land (dummy)	–0.78***	(0.20)	0.16	(0.13)	–0.43**	(0.22)	–0.14	(0.20)
Off-farm employment (dummy)	0.15	(0.22)	–0.45***	(0.14)	–0.48***	(0.19)	–0.69***	(0.25)
Remittances (dummy)	0.71***	(0.25)	–0.04	(0.15)	–0.01	(0.24)	–0.52***	(0.19)
Business (dummy)	0.04	(0.20)	–0.36***	(0.14)	–0.13	(0.23)	–0.29	(0.19)
Distance to nearest market town (Km)	0.003	(0.02)	–0.001	(0.01)	0.005	(0.02)	0.06**	(0.02)
Road type (*1=good*)	0.02	(0.20)	0.14	(0.12)	0.19	(0.21)	–0.10	(0.22)
Transport facility (dummy)	–0.11	(0.29)	0.13	(0.13)	–0.22	(0.21)	0.004	(0.21)
Total land owned (acres)	–0.08	(0.09)	–0.03	(0.02)	–0.04	(0.03)	–0.08	(0.05)
Land cultivated (acres)	0.04	(0.09)	0.09**	(0.04)	0.0005	(0.06)	0.01	(0.07)
Access to credit (dummy)	–0.08	(0.22)	0.23	(0.16)	0.46*	(0.25)	1.05***	(0.20)
Annual rainfall (mm, lagged)	0.003	(0.00)	0.001	(0.00)	–0.003	(0.00)	0.0002	(0.00)
Covariance of variation (monthly precipitation)	–0.04**	(0.02)	0.003	(0.01)	–0.05*	(0.03)	–0.01	(0.02)

To be continued...

	In (Horticultural commercialization index, HCI_1)				*Selection equation (export=1)*			
	Household fixed effects (1)		Pooled OLS (2)		Year_2005 (3)		Year_2010 (4)	
Maize price ($/Kg)	0.48	(0.90)	–0.15	(0.57)	–0.91	(0.94)	1.76**	(0.88)
Maize area (percent)	–0.37*	(0.22)	–0.004	(0.00)	–0.01**	(0.00)	–0.01	(0.01)
Extension contact (dummy)					0.13	(0.23)	0.25	(0.18)
Own TV (dummy)					–0.27	(0.20)	–0.28	(0.21)
Own Radio (dummy)					–0.04	(0.36)	–0.41	(0.42)
Own phone (dummy)					–0.16	(0.21)	0.31	(0.55)
French beans price ($/Kg)					–1.08	(0.69)	5.11**	(2.29)
Snow peas price ($/Kg)					–0.92	(1.55)	15.89***	(4.39)
Year dummy(*1=2010*)	0.27	(0.44)	–0.40	(0.28)				
IMR ($\hat{\lambda}$)	–0.62**	(0.33)	0.52	(0.42)				
IMR*Year dummy(1=2010)			–0.32	(0.43)				
District fixed effects			Yes		Yes		Yes	
Constant	3.36	(3.09)	3.61***	(1.36)	10.41**	(4.30)	–14.8***	(4.53)
Number of observations	345		345		309		309	
R-squared; Pseudo R2	0.53		0.23		0.15		0.30	
F(24,74) ; F(29, 315)	3.42***		3.17**					
LR chi2(32)(model chi)					53.7***		121.5***	
Log likelihood					–150.88		–144.263	

Note: *The dependent variable for the* ***Fixed effect and pooled OLS models*** *is the share of income from export vegetable sales (log) while that of* ***the selection equations (probit model)*** *is the adoption of export market vegetable production (dummy);* ***Significance*** *at 0.01(***), 0.05(**), and 0.1(*) probability levels.* ***Standard errors*** *are given in parenthesis.* ***Household fixed effects*** *are jointly significant in model (1) above (F (246, 74) =2.40, Prob > F =0.00).* ***Product prices*** *are taken at the division level (the next lowest administrative unit after district). The* ***inverse mills ratios*** *generated in models (3) and (4) are used to estimate the fixed effect, model (1), and the pooled OLS, model (2).*

With regard to household endowment of land and capital indicators, the total land owned is negatively but and not significantly related to both the decision to participate in the export market and the extent of commercialization. The area of cultivated land shows a positive relationship with on the extent of commercialization based on pooled OLS results; however, we do not interpret this finding as pooled OLS results may be biased by sample selection. The observation made in the field was that in some districts such as Meru, households with large farms engage in other commercial activities such as tea, coffee and dairy farming. This phenomenon could explain the negative relationship between land owned and the extent and decision to commercialize in export market horticulture. Furthermore, the returns per unit area of export crops are higher compared to other crops. For example, Minot and Ngigi (2004) estimated that French beans have over twice the gross product value per unit area as domestically consumed vegetables such as potatoes and cabbage. McCulloch and Ota (2002) and Sindi (2008) also find insignificant relationships between the total land owned and the decision to participate in horticultural commercialization.

Land quality (*fertile land* dummy) had a negative and significant relationship with the decision to commercialize and on the extent of commercialization through the export market. Low productivity of land was observed across the entire household sample (28% of the total sample reported to access fertile land, see Table 2.3). As such, most farmers are dependent on fertilizers especially those supplying to the export market. This situation could explain the unexpected sign of land quality and participation in the export market.

Livestock assets are negative and statistically significantly related to the extent of commercialization along the export market pathway. Qualitative information from the field showed that farmers are increasingly diversifying their farm income sources by engaging in dairy farming, which created competition for labour and land resources. This is largely found in the Central province. Although an empirical study is required to ascertain this hypothesis, dairy farming seems to be competing with French beans production in particular because of higher prices of milk and an increased number of milk traders in the area.

With respect to demographic factors, the age of the household head has a negative and significant relationship on both the decision for and the extent of commercialization through the export market. Similarly, household size presents negative and significant relationship with the extent of commercialization. Increase in household size have been found to increase pressure on land (von Braun and Kennedy 1994), thus reducing the volume of marketed surplus as subsistence needs become a priority over commercial activities for smallholders The prices of maize, for a staple crop in Kenya, have been increasing in recent years. As such, land and labour resources otherwise allocated to vegetable production may be diverted to maize production. In light of this, two variables related to maize production—percentage of cultivated land area allocated for maize production and maize prices at the household level were included in the model. Whereas the estimate of maize price is not significant, the area allocated to maize production exhibit a negative and significant relationship on the extent of commercialization through the export market. This finding confirms an earlier suggestion that food self-sufficiency could be a priority for market participation. The share of land allocated to maize production is however positive and statistically significantly related to the decision to commercialize in 2010.

The three indicators of non-farm income endowments included in the model—access to off-farm employment, remittances and business ownership—provide mixed results. Off-farm employment and remittances are negatively and significantly related to the choice of the export market, while business ownership is negatively but not significantly related to the choice of this market. The fixed effects parameter for remittances however is positively and significantly associated with the extent of commercialization. Based on field observations, off-farm employment and the ownership of a business may also compete for labour resources required for the production and marketing of export vegetables, while access to remittances provides an alternative source of income. Thus, households are not compelled to engage in risky farm enterprises such as the production of export marker vegetables. However, producers are engaged in commercialization, remittances could provide a source of capital to meet operation costs of the capital-intensive farm enterprises. These observations however need to be confirmed using empirical analysis.

Although institutional indicators (membership in PMOs and adoption of GlobalGap standards are not included in the above model as they are endogenous to export market participation, they were observed to play a significant role in determining the extent of commercialization.[10] Farmers that are organized in groups are more likely to enter a marketing arrangement with commercial produce buyers as they can collectively meet the required volumes of produce. From 2000 to 2005, farmers invested measures to achieve compliance with GlobalGap requirements jointly through PMOs, received training together, and also carried out internal monitoring and coordination amongst themselves, thus reducing transaction costs incurred by exporters from facilitating compliance with these requirements (Okello et al., 2009). Compliance with GlobalGap standards has been

10 The correlation coefficients between *Horticultural Commercialization Index (HCI_1_export)* and the institutional variables (group membership and Globalgap compliance) are both positive and statistically significant at 1% level.

cited in past studies as a means of enhancing horticultural product acceptability to international markets (Asfaw, 2008; Muriithi et al., 2010a). Access to credit is positively and significantly associated with the decision to commercialize. This finding is consistent with Ashraf et al. (2009), who find that credit increased participation of smallholder horticultural growers in an export-crop production program.

As expected the lagged average annual rainfall variable is positive but not significantly associated with the extent of commercialization through the export market pathway, with respect to fixed effect and pooled OLS estimates. Further, agreeing with our hypothesis, the weather risk indicator (CoV) is negative and significantly related to the extent of commercialization (FE coefficients) and the decision to commercialize (probit equation, 2005). Erratic rainfall patterns are likely to reduce the output levels and have negative impacts on the decision to produce vegetables for commercial purposes.

2.6.2 Commercialization through the domestic market pathway

Table 2.5 shows the estimates of determinants of the decision and extent of commercializing through the domestic market. The table is organized in the same way as Table 3, i.e. columns (3) and (4) display the results for the selection equation in 2005 and 2010, respectively, while the estimates for the extent of commercialization with and without household fixed effects are shown in columns (1) and (2), respectively. The F-test for the fixed effect and pooled OLS models rejects the null hypothesis that all coefficients of the variables considered including the constant are zero ($p<0.01$). Similarly, the selection models are also highly significant ($p<0.01$) based on the log-likelihood ratio test. In contrast to the results for the export market channel, the inverse Mills ratio is not statistically significant in either specification, which ameliorates our concern for potential selection bias among producers for the domestic market. Among the proportional transaction cost proxies, only distance to the nearest market town appears to influence the decision to commercialize through the domestic market pathway in a positive and statistically significant way. This may be due to the same reasons as in the selection equation for the export market: Greater proximity to market towns offers alternative income-generating activities. In line with our previous explanation, access to good roads (as perceived by the farmer) exerts a positive and statistically significant influence on the extent of commercialization when household fixed effects are not included in column (2) due to better transport facilities for the produce. Similarly to the selection equation for the export market pathway, fixed transaction costs– ownership of a TV, radio and access to extension services seem not to have any significant effect on the decision to commercialize through the domestic market pathway. However, contrasting the estimates of the export market,

Table 2.5: Determinants of vegetable commercialization through the domestic market pathway

	ln (Horticultural commercialization index, HCI_1)				*Selection equation (domestic market=1)*			
	Household fixed effects (1)		Pooled OLS (2)		Year_2005 (3)		Year_2010 (4)	
Head gender (1=male)	–0.39	(0.49)	0.04	(0.22)	0.08	(0.35)	–0.16	(0.33)
Head age (years)	–0.02	(0.01)	–0.01	(0.01)	0.01	(0.01)	–0.02*	(0.01)
Head education (years of schooling)	0.04	(0.04)	0.001	(0.02)	–0.02	(0.03)	0.01	(0.03)
Household size (adult equivalent)	0.19	(0.17)	0.04	(0.11)	–0.07	(0.16)	0.09	(0.16)
Household size (squared)	–0.01	(0.01)	–0.005	(0.01)	0.01	(0.01)	–0.002	(0.01)
Oxen ownership	1.67**	(0.76)	0.64	(0.50)	0.34	(0.71)	–1.65***	(0.64)
Agricultural implements (index)	0.01	(0.29)	0.11	(0.19)	0.45*	(0.29)	0.24	(0.25)
Livestock (index)	–2.06**	(0.99)	–1.55**	(0.61)	–0.53	(0.94)	2.26**	(1.15)
Fertile land (dummy)	–0.19	(0.23)	0.24*	(0.15)	–0.35	(0.23)	–0.07	(0.19)
Off-farm employment (dummy)	0.02	(0.22)	-0.47***	(0.14)	0.14	(0.20)	0.01	(0.25)
Remittances (dummy)	-0.08	(0.22)	-0.27	(0.15)	0.14	(0.26)	0.17	(0.19)
Business (dummy)	-0.16	(0.21)	-0.31**	(0.14)	0.12	(0.26)	0.02	(0.19)
Distance to nearest market town (Km)	-0.02	(0.02)	-0.002	(0.01)	0.06**	(0.03)	0.07**	(0.04)
Road type (*1=good*)	0.17	(0.21)	0.35**	(0.15)	-0.13	(0.21)	0.06	(0.21)
Transport facility (dummy)	0.25	(0.25)	-0.02	(0.14)	-0.03	(0.22)	-0.09	(0.20)
Total land owned (acres)	0.02	(0.05)	-0.02	(0.03)	-0.01	(0.03)	0.05	(0.04)
Land cultivated (acres)	-0.04	(0.08)	0.02	(0.05)	0.07	(0.07)	-0.07	(0.06)
Access to credit (dummy)	0.12	(0.21)	0.01	(0.16)	-0.73***	(0.22)	-0.19	(0.19)
Annual rainfall (mm, lagged)	-0.002	(0.002)	0.001	(0.001)	0.01*	(0.00)	0.01***	(0.00)
Covariance of variation (CoV, monthly precipitation)	0.08***	(0.02)	0.03***	(0.01)	0.01	(0.03)	0.01	(0.03)

	In (Horticultural commercialization index, HCI_1)				*Selection equation (domestic market=1)*			
	Household fixed effects (1)		Pooled OLS (2)		Year_2005 (3)		Year_2010 (4)	
Maize price ($/Kg)	1.16	(0.82)	0.12	(0.62)	0.69	(1.27)	-0.34	(0.82)
Maize area (percent)	-0.004	(0.01)	-0.01*	(0.004)	0.001	(0.01)	-0.01**	(0.01)
Extension contact (dummy)					0.02	(0.25)	-0.04	(0.18)
Own TV (dummy)					0.21	(0.21)	-0.11	(0.20)
Own Radio (dummy)					0.18	(0.39)	0.77*	(0.41)
Own phone (dummy)					-0.01	(0.21)	-0.46	(0.41)
Potatoes price ($/Kg)					3.39	(3.34)	-0.65	(3.16)
Cabbage price ($/Kg)					-9.58**	(4.13)	2.10***	(0.80)
Year dummy (*1=2010*)	-0.41	(0.35)	-0.14	(0.28)				
IMR ($\hat{\lambda}$)	-0.40	(0.44)	-0.41	(0.40)				
IMR*Year dummy (1=2000)			0.50	(0.45)				
District fixed effects	Yes		Yes		Yes		Yes	
Constant	1.29	(2.54)	0.699	(1.57)	-5.72	(4.51)	-4.79*	(2.59)
Number of observations	431		431		309		309	
R-squared; Pseudo R2	0.181		0.2847		0.2251		0.2332	
F(24,74); F(29, 315)	1.35*		5.5***					
LR chi2(32)(model chi)					79.09***		93.25***	
Log likelihood					-136		-153	

Note: *The dependent variable for the* ***Fixed effect and pooled OLS models*** *is the share of income from domestic market vegetable sales (log) while the dependent variables for* ***the selection equations*** *is adoption of locally consumed vegetable;.* ***Significance*** *at 0.01(***), 0.05(**), 0.1(*) probability level:* ***Standard errors*** *shown in brackets.* ***Household fixed effects*** *are jointly significant in model (1)(F(259, 147) =1.97, Prob > F =0.000).* ***Product prices*** *are taken at division level. The same inverse mills ratios generated in models (3) and (4) are used in estimating the Fixed effect model (1) and the pooled OLS model (2).*

ownership of a radio is positive and statistically significantly related to the decision to commercialize through the domestic market pathway. The price of cabbage, one of the main locally consumed vegetables, exhibits a negative and statistically significant coefficient for the decision to commercialize through the domestic market in 2005, but the opposite sign in 2010. A possible explanation for this could be related to recent increases in the prices of locally consumed vegetables due to increased demand for food in the country. This is in contrast to 2005, when the prices of export market vegetables were more stable and reliable. Discussions with exporters during the field survey revealed high levels of uncertainty about product prices in the international markets that compel them to shun providing fixed prices to producers.

The extent of commercialization through the domestic market pathway is negatively and statistically significantly influenced by livestock index, access to non-farm employment and business ownership. This is given by pooled OLS estimates; except for the livestock index estimate, which is also significant in respect to the FE model. The ownership of traction animal (oxen) is positively related to the extent of commercialization through the domestic market pathway. Similarly to the FE effects estimates for the export market pathway, the size of land owned and land cultivated have insignificant effects on the extent of commercialization through the domestic pathway. With regard to weather indicators, the weather risk proxy (CoV) exhibit unexpected positive and significant association on the extent of commercialization. The unexpected sign could be related to the production patterns of domestically marketed vegetables. Unlike vegetables supplied to the international market, which are being produced throughout the year and thus likely to be affected by the annual variation of rainfall, the planting season for locally consumed vegetables is during the rainy season, which may make them less affected by the annual variability of rainfall.

With respect to the selection equation, the age of the household head is negatively related to the decision to commercialize through the domestic market. The livestock asset index value and agricultural assets are positively and significantly related to the decision to commercialize through the domestic market pathway. Correspondingly to the decision to commercialize through the export market, annual rainfall received during the previous growing season, is positively related to the decision to supply vegetables to the domestic market.

2.6.3 Commercialization through the domestic and export markets jointly

The estimates of the determinants of the decision and extent of commercialization jointly through both domestic and export market pathways are provided in Table 2.6.

Table 2.6: Determinants of vegetable commercialization jointly through export and domestic market pathways

	ln (Horticultural commercialization index, HCI_1)				*Selection equation (joint export and domestic markets=1)*			
	Fixed effects (1)		Pooled OLS (2)		Year_2005 (3)		Year_2010 (4)	
Head gender (1=male)	1.12**	(0.48)	0.20	(0.19)	–0.15	(0.30)	0.27	(0.37)
Head age (years)	–0.004	(0.02)	–0.003	(0.005)	–0.01*	(0.01)	–0.01	(0.01)
Head education (years of schooling)	–0.09*	(0.06)	0.004	(0.02)	–0.02	(0.03)	0.003	(0.03)
Household size (adult equivalent)	–0.47**	(0.20)	–0.12	(0.10)	0.05	(0.15)	0.03	(0.18)
Household size (squared)	0.03**	(0.01)	0.01	(0.01)	–0.003	(0.01)	0.004	(0.01)
Oxen ownership	1.95**	(0.88)	0.93**	(0.38)	0.12	(0.65)	–0.83	(0.65)
Agricultural implements (index)	0.82***	(0.32)	0.11	(0.17)	0.47*	(0.25)	0.16	(0.29)
Livestock (index)	–1.50	(1.49)	–2.56***	(0.55)	0.24	(0.84)	1.12	(0.94)
Fertile land (dummy)	–0.65**	(0.29)	0.001	(0.14)	–0.67***	(0.21)	–0.31	(0.22)
Off-farm employment (dummy)	0.45*	(0.25)	–0.24**	(0.12)	–0.26	(0.18)	–0.61**	(0.27)
Remittances (dummy)	0.29	(0.30)	0.04	(0.14)	0.05	(0.23)	–0.17	(0.21)
Business (dummy)	0.05	(0.22)	–0.49***	(0.13)	–0.06	(0.23)	–0.29	(0.20)
Distance to nearest market town (Km)	–0.02	(0.02)	0.01	(0.01)	0.03	(0.02)	0.07***	(0.02)
Road type (*1=good*)	0.41	(0.47)	0.13	(0.13)	–0.07	(0.19)	0.20	(0.26)
Transport facility (dummy)	–0.12	(0.41)	0.07	(0.11)	–0.16	(0.19)	0.02	(0.22)
Total land owned (acres)	–0.06	(0.11)	–0.02	(0.03)	–0.05	(0.04)	–0.01	(0.06)
Land cultivated (acres)	0.05	(0.18)	0.05	(0.05)	0.08	(0.08)	0.05	(0.07)
Access to credit (dummy)	0.59	(0.45)	–0.06	(0.15)	–0.34	(0.21)	0.69***	(0.22)
Annual rainfall (mm, lagged)	–0.001	(0.003)	0.0004	(0.001)	0.002	(0.003)	0.002	(0.002)
Covariance of variation ((CoV, monthly precipitation)	–0.04*	(0.02)	0.01	(0.01)	–0.04	(0.03)	0.01	(0.03)
Maize price ($/Kg)	1.71	(1.62)	–0.15	(0.48)	–0.46	(0.96)	1.60*	(0.95)

To be continued...

	In (Horticultural commercialization index, HCI_1)				*Selection equation (joint export and domestic markets=1)*			
	Fixed effects (1)		Pooled OLS (2)		Year_2005 (3)		Year_2010 (4)	
Maize area (percent)	–0.01	(0.01)	–0.003	(0.003)	–0.01*	(0.005)	–0.01	(0.01)
Extension contact (dummy)					0.02	(0.22)	0.12	(0.20)
Own TV (dummy)					–0.04	(0.19)	–0.19	(0.22)
Own Radio (dummy)					0.14	(0.34)	–0.27	(0.42)
Own phone (dummy)					–0.14	(0.20)	–0.28	(0.58)
Potatoes price ($/Kg)					4.20	(3.77)	–2.82	(3.55)
Cabbage price ($/Kg)					–11.03***	(4.21)	3.34**	(1.65)
French beans price ($/Kg)					–0.49	(0.69)	11.84***	(4.79)
Snow peas price ($/Kg)					0.30	(1.98)	21.58***	(8.48)
Year dummy(*1=2010*)	–0.36	(0.77)	–0.33	(0.40)				
IMR ($\hat{\lambda}$)	–0.70	(0.45)	–0.26	(0.26)				
IMR*Year dummy (1=2000)			0.32	(0.36)				
District fixed effects			Yes		Yes		Yes	
Constant	7.43*	(3.89)	4.02***	(1.40)	2.93	(4.17)	–25.24	(8.89)
Number of observations	223		223		309		309	
R-squared/Pseudo R2	0.60		0.37		0.17		0.24	
F(24,178)/F (29, 193)	11.19***		3.9***					
LR chi2(32)(Model Chi-square)					74.1***		80.1***	
Log likelihood					–177.1		–126.4	

Note: *The dependent variable for the* ***Fixed effects and pooled OLS equations*** *is the share of income from both domestic and export market vegetable sales (log), while the dependent variables for* ***the selection equation*** *is supplying jointly to local (domestic) and export markets;* ***Significance*** *at 0.01(***), 0.05(**), 0.1(*) probability level.* ***Standard errors*** *are shown in brackets.* ***Household fixed effects*** *are jointly significance in the fixed effect model (1)* (F(259, 147)=1.97, Prob > F =0.00). ***Product prices*** *are taken at division level. The same inverse mills ratios generated in models (3) and (4) are used in estimating the Fixed effect model (1) and the pooled OLS model (2).*

Unlike the previous discussion, the dependent variable for the selection model given in Table 2.6 is a dummy defined as 1 if a household produced vegetables for both export and domestic markets. In the same way as the previous estimates, all models are significant (*$p<0.01$*). The inverse mills ratios are insignificant; hence, there is no evidence of selection bias in the decision to participate in both market pathways.

Similarly to the domestic market pathway, proportional transaction costs did not seem to influence the extent of commercialization jointly through both markets, except the distance to the nearest market which is positively and significantly related to the decision to participate in both markets. Similarly to the domestic and export market pathways separately, all fixed transaction cost indicators are insignificant except the vegetable prices. The divisional level prices of cabbages, French beans and snow peas are positive and significantly related to the decision to participate in both markets in year 2010. In contrast to 2010, the divisional level prices of cabbage show a negative and significant influence on the decision to commercialize jointly through both markets in 2005.

In the same way as those supplying the domestic market, the agricultural asset index is positively and significantly associated with the choice of both export and domestic markets, while diversification of income through livestock ownership decreased the extent of commercialization through both markets. The share of cultivated area allocated to maize was negatively and significantly associated within the decision to commercialize through both markets in 2005. The estimate of this variable in 2010 shows the same sign but is not significant. This finding sheds some light on the competition of food crops and marketed crops for further and in-depth research.

Correspondingly to the domestic market pathway, household size is negatively and significantly associated with the extent of commercialization through both markets. Similarly to participation in the export market, education of the household head is negatively and significantly related to the extent of commercialization through both domestic and export markets.

The institutional indicators—membership in a PMO and GlobalGap compliance, which are not included in the model due to endogeity, are likely to have positive relationships with the extent of commercialization through both markets. The possibility of spill-overs effects from GlobalGap training to the production of domestic market vegetables is legitimate, however, further research is required to ascertain this hypothesis. PMOs are associated with the production of export crops, but the modern domestic market channels also dictate the need for collective action in the form of producer groups/organizations to reduce the transaction costs of trading with several scattered smallholders. However, such initiatives are still limited and mainly concentrated in regions near urban areas (Rao & Qaim, 2011). It was observed from our survey area that farmer groups engaged in the production of locally consumed vegetables do not exist.

2.6.4 Dynamics of smallholder participation in the commercialization of horticulture

The multinomial logit regression results are reported in Table 2.7. As described in section 2.5.2, households are classified into five groups (export market suppliers, domestic market suppliers, joint market suppliers, market switching suppliers and exitors). The "exitors" group is used as the comparison group. Differences are evaluated between the four groups of households that are still producing and the commercial market exitors. This analysis deviates from the previous analysis as it enables a comparison of the different groups of households with regard to participation in different vegetable market pathways.

In terms of farmer characteristics, exitors are significantly older than export market suppliers and suppliers of both markets, and have significantly larger households than domestic market suppliers. The other coefficient estimates on household size and age of household head have similar positive relationships signs but are insignificant. This finding agrees with the estimations given in the previous section. In terms of assets, exitors have significantly fewer livestock assets than respondents who maintained either their market affiliations or the switched market pathway, and fewer agricultural assets than the joint market group. This suggests that livestock production may not be competing with commercial vegetable production, but could be an income diversification strategy for small producers. Exitors have a significantly lower probability of accessing off-farm employment than the domestic and switched market groups. This suggests that exitors are engaged in alternative farm enterprises other than vegetable production.

With respect to land, exitors own significantly less land than domestic market suppliers. The exitors own significantly larger cultivated areas than the domestic and switched markets households. The coefficient for export market households was negative but not significant. This observation can be explained from two possible perspectives. On one hand, farmers producing commercial crops such as tea and coffee, who could be part of those exiting the vegetable market, own relatively large pieces of cultivated land (see Appendix 2B). On the other hand, farmers may have reallocated land resources to maize production whose supply have dwindled have over the last few years in Kenya. The coefficients of the share of land cultivated allocated to maize production are negative for all groups relative to the exitors. The coefficients of this variable for joint and switched market household are negative and statistically significant.

Joint and domestic households have significantly greater remittances than exitors. In terms of infrastructure, exitors are more likely to have better road networks. This finding highlights the importance of physical infrastructure in offering

Table 2.7: Multinomial logit results for vegetable Commercialization pathways

	Export market suppliers		Domestic market suppliers		Both market suppliers		Switched markets	
Head gender (1=male)	0.88	(1.16)	1.46	(1.04)	0.27	(0.77)	1.36**	(0.70)
Head age (years)	–0.09*	(0.05)	–0.01	(0.02)	–0.08***	(0.03)	–0.02	(0.02)
Head education (years of schooling)	–0.09	(0.11)	–0.03	(0.09)	–0.10	(0.09)	0.04	(0.07)
Household size (adult equivalent)	–0.48	(0.56)	–0.81*	(0.45)	–0.26	(0.52)	–0.21	(0.39)
Household size (squared)	0.05	(0.04)	0.06*	(0.04)	0.00	(0.04)	0.00	(0.03)
Oxen ownership	–1.08	(2.60)	–0.30	(1.78)	–0.71	(1.91)	–1.38	(1.56)
Agricultural implements (index)	–0.32	(1.17)	1.02	(0.71)	1.59*	(0.86)	0.89	(0.62)
Livestock (index)	8.50*	(4.57)	4.97*	(2.76)	4.86*	(2.69)	4.45**	(2.25)
Fertile land (dummy)	0.41	(1.02)	–0.41	(0.58)	–1.7***	(0.66)	–0.90**	(0.46)
Off-farm employment (dummy)	–1.93	(1.45)	1.58*	(0.57)	0.66	(0.56)	0.72*	(0.44)
Remittances (dummy)	0.75	(1.10)	1.46*	(0.77)	1.67**	(0.85)	0.37	(0.69)
Business (dummy)	–1.87	(1.33)	0.43	(0.64)	–0.59	(0.82)	0.11	(0.52)
Distance to nearest market town (*Km*)	–0.33	(0.35)	–0.05	(0.08)	–0.07	(0.07)	–0.01	(0.06)
Road type (*1=good*)	–0.73	(0.83)	–1.12*	(0.59)	–1.68***	(0.63)	–1.03**	(0.44)
Transport facility (dummy)	–1.38*	(0.82)	–0.27	(0.59)	–0.71	(0.63)	–0.42	(0.49)
Total land owned (acres)	–0.33	(0.26)	0.12*	(0.06)	–0.06	(0.12)	0.05	(0.06)
Land cultivated (acres)	–0.06	(0.50)	–0.64***	(0.22)	0.22	(0.21)	–0.27*	(0.15)
Access to credit (dummy)	1.29*	(0.85)	–0.98	(0.81)	0.09	(0.73)	0.02	(0.56)
Annual rainfall (mm, lagged)	0.01	(0.01)	0.01	(0.01)	0.02*	(0.01)	0.00	(0.01)

To be continued...

	Export market suppliers		Domestic market suppliers		Both market suppliers		Switched markets	
Covariance of variation (CoV, monthly precipitation)	–0.09	(0.17)	0.05	(0.08)	–0.04	(0.08)	–0.02	(0.06)
Maize price ($/Kg)	–3.86	(4.10)	1.18	(2.34)	–1.90	(3.11)	1.02	(2.07)
Maize area (%)	–0.01	(0.03)	–0.02	(0.01)	–0.03*	(0.01)	–0.02**	(0.01)
Extension contact (dummy)	0.20	(1.19)	–0.48	(0.72)	–0.65	(0.66)	–0.40	(0.55)
Own TV (dummy)	0.50	(1.00)	1.06*	(0.57)	1.07*	(0.65)	0.56	(0.46)
Own Radio (dummy)	0.85	(1.69)	–0.81	(1.04)	–1.25	(1.07)	–1.37*	(0.78)
Own phone (dummy)	–0.87	(0.91)	–0.04	(0.66)	–0.90	(0.65)	–0.43	(0.44)
Potatoes price ($/Kg)	10.12	(21.93)	–7.33	(10.69)	1.65	(11.33)	–6.74	(8.51)
Cabbage price ($/Kg)	4.00	(18.08)	–16.29	(13.48)	–24.05*	(13.55)	–5.10	(9.39)
French beans price ($/Kg)	–4.35*	(2.53)	2.48	(1.89)	–5.3***	(1.85)	–0.89	(1.56)
Snow peas price ($/Kg)	–17.71	(17.11)	5.73	(5.93)	3.12	(5.60)	3.25	(4.35)
Constant	11.91	(28.11)	–10.08	(14.07)	–8.13	(15.46)	1.69	(10.63)
Number of observation	306							
Wald chi2(136)	3460***							
Pseudo R2	0.30							
Log pseudo likelihood	–312.7							

Note: Categories of farmers in 2010 are based on 2005 characteristics; "exitors" is the comparison group. Robust standard errors are in parentheses. No new entrants were identified. ***Significance*** *at 0.01(***), 0.05(**), 0.1(*) probability level. The number of observations excludes 3 households who were non-sellers in 2005.*

alternative rural livelihoods other than vegetable production for households who are constrained to participate in vegetable markets. Although the coefficients are not statistically significant, lagged rainfall estimates for the households participating in the production of vegetables for sale are positive while those of the coefficient of variation (CoV) are negative (except for those who maintained domestic market). This supports our hypothesis that higher annual rainfall in a particular year may positively influence the decision of farmers to participate in the commercialization of horticulture in the subsequent year. On the other hand, a higher CoV implies an erratic distribution of rainfall across the year, which is likely to have negative influence the decision of farmers to participate in the market. Most fixed transaction cost indicators are not significantly different between exitors and participants under different market participation categories, except ownership of a TV and radio. Exitors are less likely to own a TV compared to domestic market suppliers and those supplying both markets, while they are more likely to own a radio than those households that switched markets.

2.7 Conclusions and policy implications

This study has described the forms of market pathways for the commercialization of smallholder vegetable farming and looked at the dynamics of smallholders' market participation. It has also empirically analyzed factors influencing households' decisions on the commercialization of vegetables through different market pathways. By using two-wave data collected from same households, this study has been able to capture market behavior of smallholder horticultural farmers that might vary across time. In addition to addressing horticultural market pathways separately, the study utilizes panel data methods that accounts for endogenity.

Overall, the number of smallholders participating in the production of vegetables for commercial purposes has been decreasing. For instance, a fifth of those who participated in the trading of vegetables in 2005 had exited by 2010 and the decline is greater among producers for the export than for the domestic market. Some of the responsible factors mentioned by the farmers themselves are weather risks, low productivity, high inputs costs, and unstable international market prices for vegetables. The standards imposed by international markets also present a challenge to resource-constrained small-scale producers.

The regression model results show that the age of the household head and household size are negatively related to vegetable commercialization, which may be related to risk aversion or less flexibility to adopt new techniques among farmers who are older or have a large number of dependants. Alternative farm enterprises such as livestock farming, off-farm and business activities appear to be competing with commercialization of horticulture across the export and domestic market pathways. Further, production of staple crops (maize in this case) seems to be competing for land resources as revealed by the negative relationship between area allocated to maize and commercialization decisions. Although results reveal mixed findings regarding the role of transaction costs on the decision and extent of commercialization, it is clear that rural farms located further away from market towns are more likely to commercialize since they have limited income diversification opportunities. However, they are conversely constrained by poor rural roads especially during high rainfall seasons. Focusing at each market pathway, credit and high vegetable prices show a positive relationship with export market participation, while the extent of commercialization is positively associated with remittances and the size of cultivated land. The decision to participate in the domestic market, on the other hand, is positively associated with livestock ownership, annual rainfall, longer distances to the nearest market town, and ownership of a radio. However, due to the competition for labor and land resources, a negative relationship exists between livestock ownership and the extent of commercialization through the domestic market. Commercialization through both the domestic and export markets jointly is found to be positively associated with credit, longer distances to the nearest market town, and high vegetable prices. Although not included in the empirical analysis due to endonegenity, membership in PMOs and GlobalGap compliance appears to impact positively on the level of commercialization. Multinomial logit results indicate that exiting farmers have significantly older household heads, own less assets (agricultural and livestock), and own less productive land than farmers who continued to engage in commercial vegetable production.

Precipitation variables are included in the regression models to incorporate the negative weather risks to smallholder market participation. The findings reveal that total average annual rainfall during a certain period of time impacts positively on decisions to participate in domestic market in the subsequent period. On the other hand, rainfall variability or weather risk, measured as coefficient of variance (CoV) using monthly rainfall data for each year of survey is negatively related to the choice and extent of commercialization of vegetables except for domestic market. Whereas vegetables supplied to the international market in most cases are produced throughout the year, domestic market vegetables are dependent on rainfall and thus timed for the rainy season. As such, export vegetables are more likely to be affected by high variability of annual rainfall than domestic vegetables.

A number of policy recommendations can be derived from the above discussion. Driven by high population growth and rural-urban migration, domestic market vegetables and other locally consumed farm produce, e.g. milk and maize, present market potential for small producers. Taking advantage of this opportunity, however, requires physical infrastructure and some form of institutional support. The initiative of the government to provide subsidized fertilizer to small farmers established in 2011, for instance, is a start in this direction. However, the government needs to explore long-term strategies for production of fertilizer locally and avoid reliance on imported fertilizers. Furthermore, there is need to explore strategies to adapt to unfavourable weather conditions and, in order to maintain their position in the export market, small producers require up-to-date information services to stay on top of evolving market requirements. Policy inventions to promote capacity building of the existing extension personnel are required so that they can adequately advise farmers on how to overcome the evolving production and marketing challenges. Establishment of sustainable farmer groups as well as support to compliance with private market requirements such GlobalGap standards and exploring the adoption of equivalent local guidelines such as KenyaGap remains of policy relevance for small horticultural producers.

Further research is recommended on the effects of national and global food price volatility on the commercialization of vegetables. Similarly, an explicit investigation of the role of producer and marketing organizations and GlobalGap standards for the extent of commercialization is beyond the scope of this study but worth further attention. Further research is also suggested to explore the role of transaction costs by quantifying in monetary terms the costs and benefits involved under each market pathway.

Appendix 2A: French bean export volumes

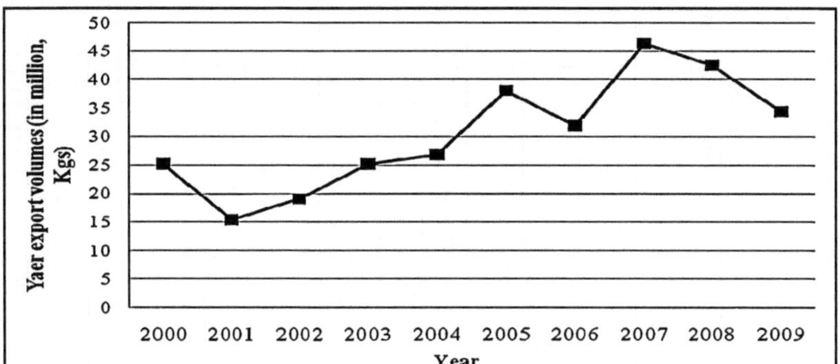

Source: Based on HCDA Export Statistics (2000–2009)

Appendix 2B: Characteristics of households by vegetable output market

	Total (n=618)		Export market suppliers (E) (n=122)		Domestic market suppliers (D) (n=208)		t-test (E-D)	Both markets suppliers (B) (n=223)		t-test (E-B)	t-test (D-B)	Non-sellers (NS) (n=65)		t-test (E-NS)	t-test (D-NS)	t-test (E-NS)
Variables	**Mean**	**SD**	**Mean**	**SD**	**Mean**	**SD**	**diff.**	**Mean**	**SD**	**diff.**	**diff.**	**Mean**	**SD**	**diff.**	**diff.**	**diff.**
Household head gender (1=male)	0.90	0.30	1.00	0.20	0.90	0.30	*0.05*	0.90	0.30	*0.04*	*–0.01*	0.83	0.38	*0.12****	*0.07*	*0.08**
Household hold head age (years)	48.70	12.50	45.00	11.20	50.74	12.53	*–5.7****	46.90	12.00	*–1.85*	*3.87****	55.4	12.8	*–10.3****	*–4.6****	*–8.5****
Head education (years of schooling)	8.70	3.50	8.70	2.90	8.62	3.95	*0.08*	8.90	3.20	*–0.17*	*–0.25*	8.31	3.68	*0.40*	*0.31*	*0.56*
Household size (adult equivalent)	4.90	2.20	4.70	2.10	5.09	2.32	*–0.35*	4.90	2.20	*–0.13*	*0.22*	4.75	2.25	*–0.01*	*0.34*	*0.13*
Oxen ownership (dummy)	0.00	0.20	0.10	0.20	0.02	0.13	*0.04***	0.00	0.20	*0.04**	*–0.01*	0.04	0.20	*0.02*	*–0.03*	*–0.02*
Agricultural implements (index)	1.00	0.40	1.00	0.40	1.07	0.41	*–0.04*	1.10	0.40	*–0.04*	*–0.01*	0.84	0.43	*0.19****	*0.22****	*0.23****
Livestock (index)	0.10	0.10	0.10	0.20	0.14	0.16	*0*	0.10	0.10	*0*	*0*	0.11	0.11	*0.03*	*0.03*	*0.03*
Fertile land (dummy)	0.30	0.40	0.40	0.50	0.31	0.46	*0.09**	0.20	0.40	*0.20****	*0.11****	0.22	0.41	*0.19****	*0.10*	*–0.01*
Off–farm employment (dummy)	0.30	0.50	0.30	0.40	0.33	0.47	*–0.06*	0.30	0.50	*–0.07*	*–0.01*	0.20	0.40	*0.07*	*0.13***	*0.14***
Remittances (dummy)	0.30	0.40	0.20	0.40	0.32	0.47	*–0.08*	0.20	0.40	*0.04*	*0.12*	0.45	0.50	*–0.21****	*–0.13***	*–0.25****
Business (dummy)	0.30	0.50	0.20	0.40	0.34	0.47	*–0.1*	0.20	0.40	*0*	*0.09*	0.40	0.49	*–0.16***	*–0.06*	*–0.16****
Distance to nearest market town (Km)	4.00	4.60	2.80	3.30	3.93	4.10	*–1.1*	5.20	5.60	*–2.34*	*–1.25*	1.97	2.48	*0.87**	*2.00****	*3.21****
Road type (1=good)	0.50	0.50	0.50	0.50	0.55	0.50	*–0.07*	0.40	0.50	*0.04*	*0.11*	0.65	0.48	*–.17***	*–0.10*	*–0.21****
Total land owned (acres)	3.10	4.40	2.90	5.50	3.49	5.26	*–0.57*	2.80	2.70	*0.13*	*0.7*	3.67	4.28	*–0.75*	*–0.18*	*–0.88***
Land cultivated (acres)	2.20	2.10	2.10	1.90	2.21	2.61	*–0.1*	2.20	1.70	*–0.08*	*0.02*	2.55	1.97	*–0.44*	*–0.34*	*–0.36*
Access to credit (dummy)	0.40	0.50	0.60	0.50	0.31	0.46	*0.26*	0.40	0.50	*0.21*	*–0.05*	0.34	0.48	*0.23****	*–0.03*	*0.02*
Annual rainfall (mm, lagged)	1073.7	116.7	1084.1	123.2	1059.0	111.8		1109.7	89.4			977.4	140.2			

	Total (n=618)		Export market suppliers (E) (n=122)		Domestic market suppliers (D) (n=208)		t-test (E-D)	Both markets suppliers (B) (n=223)		t-test (E-B)	t-test (D-B)	Non-sellers (NS) (n=65)		t-test (E-NS)	t-test (D-NS)	t-test (E-NS)
Variables	**Mean**	**SD**	**Mean**	**SD**	**Mean**	**SD**	**diff.**	**Mean**	**SD**	**diff.**	**diff.**	**Mean**	**SD**	**diff.**	**diff.**	**diff.**
Covariance of variation (monthly precipitation)	54.20	8.00	57.30	9.30	53.63	7.38		52.10	7.20			57.61	6.87			
Vegetable contact (dummy)	0.50	0.50	0.80	0.40				0.80	0.40	*0.06*						
Member farmer group (dummy)	0.40	0.50	0.60	0.50				0.70	0.50	*–0.05*						
Member farmer group (years)	1.80	3.80	3.40	5.20				2.10	3.40	*1.28*						
GlobalGap compliant (dummy)	0.30	0.50	0.60	0.50				0.50	0.50	*0.07*						
Extension contact (dummy)	0.50	0.50	0.70	0.40	0.55	0.50	*0.18*	0.70	0.50	*0.04*	*–0.14*	0.32	0.47	*0.41****	*0.23****	*0.37****
Transport facility (dummy)	0.60	0.50	0.60	0.50	0.55	0.50	*0.04*	0.50	0.50	*0.11*	*0.08*	0.52	0.50	*0.07*	*0.03*	*–0.05*
Own TV (dummy)	0.50	0.50	0.40	0.50	0.53	0.50	*–0.11*	0.50	0.50	*–0.04*	*0.06*	0.52	0.50	*–0.10*	*0.01*	*–0.05*
Own Radio (dummy)	0.90	0.20	0.90	0.30	0.97	0.18	*–0.04*	0.90	0.30	*–0.01*	*0.03*	0.91	0.29	*0.02*	*0.06***	*0.03*
Own phone (dummy)	0.80	0.40	0.70	0.50	0.84	0.37	*–0.13*	0.70	0.40	*–0.04*	*0.09*	0.89	0.31	*–0.19****	*–0.06*	*–0.15****
French beans price ($/Kg)	0.50	0.20	0.50	0.20	0.47	0.15	*0.02*	0.50	0.20	*–0.03*	*–0.05*	0.42	0.08			
Snow peas price ($/Kg)	0.70	0.10	0.70	0.10	0.74	0.10	*0*	0.80	0.10	*–0.02*	*–0.02*	0.67	0.07			
Potato price ($/Kg)	0.20	0.10	0.20	0.10	0.17	0.07	*0.04*	0.20	0.10	*0*	*–0.04*	0.15	0.04			
Cabbage price ($/Kg)	0.2	0.2	0.1	0.1	0.22	0.22	*–0.1*	0.20	0.20	*–0.05*	*0.05*	0.11	0.11			
Maize price ($/Kg)	0.26	0.07	0.24	0.09	0.26	0.01	*0.02**	0.24	0.01	*–0.01*	*0.01*	0.32	0.15	*–0.08****	*–0.06****	*–0.08****
Maize area (%)	21.86	18.67	19.50	19.49	24.46	18.68	*–4.95***	19.09	17.80	*0.42*	*5.37****	27.62	18.0	*–8.11****	*–3.16*	*–8.53****
Potato area (%)	8.23	15.48	1.81	5.85	11.87	19.31	*–10****	9.78	14.92	*–7.97****	*2.09*	3.10	10.3	*–1.29*	*8.76****	*6.67****

NB: Significance at 0.01(***), 0.05(**), 0.1(*) probability level.

Chapter Three

3. Does Commercialization of Smallholder Horticulture Reduce Rural Poverty? Evidence from Household Panel Data in Kenya

3.1 Introduction

Commercialization of smallholder agriculture is viewed as a pathway towards economic growth and development in many developing countries that have agriculture dependent economies (von Braun et al., 1994; Pingali & Rosegrant, 1995; Pingali, 2007). Commercialization and the diversification of horticulture, especially of high-value crops, have been identified in the past two decades as one of the fastest growing agricultural sub-sectors in Sub-Saharan Africa (Gioè, 2006; Afari-Sefa, 2007; Henson & Jaffee, 2008). Until 2008, horticultural exports have been growing at an impressive rate in Kenya (Figure 1.1). As a result of this remarkable growth, policy makers, donors, and researchers perceive the rapid growth of the sub-sector as a viable "pro-poor" rural development strategy, assuming that the growth is reaching rural smallholders, the majority of whom are involved in horticulture. However, there are concerns regarding the perceived impacts of horticulture on rural smallholders.

The first concern is that smallholder farmers are being forced out of the horticultural business. With the increasing integration of developing countries in global market trade, there are emerging forms of non-tariff barriers such as food quality and safety standards for which compliance may be cost prohibitive to small producers (Dolan & Humphrey, 2000; Jaffee, 2003; Henson & Reardon, 2005; Jaffee et al., 2005; Okello, 2005; Muriithi et al., 2010a). As a result, past studies have found that exporters are increasingly shifting away from smallholder contract farming to large-scale producers or diversifying their own agribusiness into crop production (Graffham et al., 2007; Okello, et al., 2007; Maertens & Swinnen, 2009). The modern vegetable marketing channels in the domestic market also seem to be unfavourable to smallholder farmers (Neven & Reardon, 2004). Similar to the international market, domestic supermarkets have established food safety standards (Neven et al., 2009). The alternative market pathways for smallholders are traditional supply chains that are highly uncoordinated and provide the least returns (Muendo & Tschirley, 2004).

Also of concern are the existing and emerging challenges faced by smallholder horticultural producers. These include: the lack of physical infrastructure (information technology, roads, and markets), high marketing risks and transaction costs, the lack of access to credit and production technologies such as high yield crop varieties, affordable fertilizer, post-harvest processing equipment, irrigation infrastructure, and high production costs (Jaffee, 2003; Adekunle et al., 2012). There is also emerging competition in the international market, especially from North Africa and South America, where the horticultural production costs may be lower due to subsidized farm inputs (Adekunle et al., 2012). Furthermore, an increasing trend in the vertical integration of production, processing and packaging among agribusinesses seems to be replacing the role of small producers (Adekunle et al., 2012). Recently a high influx of horticultural produce into the regional markets has been observed from Kenya's neighbours such Tanzania and Uganda, where production costs are lower (USAID, 2011; GoK, 2012). The high production cost in Kenya are partially related to reliance on imported inorganic fertilizers whose prices have risen sharply over the last decade (Gitau et al., 2012), as well as increasing labour costs due to increasing inflation. Smallholders that dominated the market for many years are being replaced with larger-scale agribusinesses linked to exporters that manage such integrations (Adekunle et al., 2012). Despite the growing demand for horticultural produce, smallholder market participation continues to decrease. Therefore, gains derived from horticultural commercialization may be constrained by these challenges, especially among the smallholder farmers who have limited ability to deal with rising agricultural production costs and marketing related risks and increasing the uncertainty of horticulture as a rural development strategy.

The perception of smallholder horticultural commercialization as a means of reducing poverty at the micro-level is often based on cross-sectional, which report the economic status of farmers at the time of the survey and typically focus on export-oriented market participation (for example McCulloch & Ota, 2002; Asfaw, 2008; Maertens & Swinnen, 2009). Empirical studies that measure the extent of horticultural sub-sector has impacts on rural poverty using panel data are non-existent. To examine the role of horticultural farming as a pro-poor strategy of rural development, an estimation of the medium-term or long-term livelihood impacts of smallholder commercialization is of policy relevance. This is especially important in Kenya where, although agriculture is the backbone of the economy and mainly concentrated in rural areas, poverty is significant with about 49.1% of the rural population living under the poverty line (IMF, 2012). The extent to which long or medium-term economic gains from smallholder commercialization reach the rural poor and whether the sub-sector is an effective pro-poor strategy in Kenya remains debatable.

This study will contribute to the existing literature on the impact of commercialization of smallholder horticulture by assessing household welfare from both income and asset ownership perspectives in Kenya.[11] To overcome the challenges for cross-sectional data in evaluating the impacts of horticultural farming, this study uses two-wave household-level panel data spanning five years (2005–2010). Moreover, the study triangulates the analysis by using methods that exploit the panel nature of the data to account for observed and unobserved heterogeneity in the sample, thus improving upon methods that make no such considerations or that relied exclusively on cross-sectional data methods. First, we applied a naïve (pooled) ordinary least square method that does not account for endogeneity. Second, a standard fixed effect approach that accounts for time invariant household effects is considered. Third, a two-step fixed effect approach is applied to formulate a panel data model similar to two-step Heckman's (1976) framework. In addition to accounting for time invariant household effects, the two-step procedure can address self-selection bias by the inclusion of a correction term in the poverty outcome model generated using a commercialization (selection) model based on observable variables. Further, we attempt a fixed effect instrumental variable approach that account for potential reverse causality.

3.2 Literature review

Two strands of literature are reviewed. First is a review on the poverty impacts of smallholder commercialization of horticultural farming and second, various methods for assessing those impacts. The purpose of the review is to inform the

11 Most studies prefer using household expenditure or consumption to income because expenditure is less susceptible to seasonal and yearly fluctuations thus generate less measurement errors (Deaton, 1989, 1991). However some past studies that used both income and consumption do not find consumption to be superior to income as an indicator of economic welfare (see Deaton 1997 & Field 2003 in Suri et al. (2009). Suri et al. (2009) also compare an analysis done in Kenya using expenditure data and income data by two different institutions- the Central Bureau of Statistics and the Tegemeo Institute of Agricultural Policy and development respectively. The results from the two analyses provide similar findings regarding the welfare position of the households. Minot and Roy (2007) also recommend estimating poverty based on income rather than expenditure while studying the impact of high-value agriculture such as horticulture since the research interest is mainly on land and labour substitution effects. However this study combines income and asset measurements which together provide more robust results on the well-being of a household.

conceptual framework and select the empirical models to guide the analysis of smallholder commercialization and the aforementioned poverty effects.

The impact of the smallholder commercialization of agriculture can be classified into three categories following Moti et al. (2009). First-order impacts are mainly income and employment effects that are directly reflected in household welfare. Following this classification, several studies on the commercialization of agriculture have mainly concentrated on first order effects (for example von Braun & Immink, 1994; Dorsey, 1999; McCulloch & Ota, 2002; Minot & Ngigi, 2004; Goletti, 2005; Sindi, 2008; Carletto et al. 2009). The second-order effects include gender, health, nutrition and assets and are usually dependent on the level of income attained through the commercialization (Kennedy, 1994; Niemeijer & Hoorweg, 1994; Fischer & Qaim, 2012a). Third-order effects are micro-level impacts beyond the household level, such as environmental effects (Seshan, 2005).

On one hand, smallholder commercialization provides direct and indirect income benefits to rural households and thus improves their livelihoods (von Braun & Kennedy, 1994). On the other hand, the transformation of subsistence agriculture to commercial enterprises may have adverse consequences, such as the presence of market price distortions as in Barrett and Dorosh (1996), or due to reallocation of resources from food to commercial crops as in Klasen (2006).

Studies that have analyzed poverty effects of smallholder horticultural farming by examining income generated either from labour or the product market (or expenditure as a proxy for income), have used different methods to measure the effects of horticultural farming. These methods include; gross margin analysis (von Braun & Immink, 1994; Minot & Ngigi 2004; Rao & Qaim 2011), comparison of mean incomes of those participating in the horticultural commercialization and non-participants; and the use of ANOVA or regression analysis on the income of participants and non-participants or using matched pairs of the same utilizing cross-sectional data (von Braun et al., 1991). Gross margin analysis has been criticized, as it cannot infer on poverty impacts since it does not measure the effects on total income (Minot & Roy, 2007). The comparison of mean income methods is unable to establish the direction of causality, while ANOVA does not control for factors such as land size and the use of irrigation, and is thus likely to produce biased estimates due to selection bias (Minot & Roy, 2007). Another set of methods are regression methods that control for various factors that may favour participation in horticulture for some farmers but not for others (McCulloch & Ota 2002; Asfaw, 2008; Maertens & Swinnen, 2009; and Afari-Sefa, 2007), however most of these studies use cross-sectional data. To our knowledge, no prior study in Kenya has used panel data to analyze smallholder horticultural commercialization effects on farm household income and simultaneously distinguished the effects of different market pathways.

Panel data methods can produce better estimates of the income effects of commercialization since they are able to control for both observed and unobserved time-invariant household characteristics. A few exceptions of studies that utilized panel data methods to measure the impact of horticulture include von Braun and Immink (1994) and Carletto et al., (2009). Von Braun and Immink (1994) find positive impact of export horticulture among small-scale farmers in Guatemala using a two-wave panel dataset (1983–1985) and expenditure as a proxy for income. Carletto et al. (2009) study small-scale export crops producers in the same country using panel data spanning over 20 years (1985–2005). The findings of Carletto et al. (2009) contradict the conclusions of other cross-sectional data studies in that although the consumption status (as a proxy for income) of all households improved, the extent of improvement among long-term market participants was less than the changes experienced by non-participants. Similarly, using durable assets and housing conditions dimensions of poverty, households that adopted commercialization early and withdrew from snow pea production by early 1990 and increased their off-farm income portfolios, demonstrated the greatest improvements compared to farmers that continued to participate and non-participants. In Kenya, most cross-sectional data studies have qualified smallholder horticulture as "pro-poor". Medium-term as well as long-term impact studies of commercial horticultural farming do not exist. Although income or expenditure measures of poverty are simple to calculate and easy to compare across regions, countries, and time periods, they have been challenged for undermining the aspects of poverty beyond observed income and expenditure measurements (Alkire & Santos, 2010). Asset- based poverty measurements on the other hand are commended for their ability to capture non-monetary facets of poverty and thus are suitable for measuring structural and stochastic poverty transitions better than income snapshots because assets are accumulated over time and persist longer (Moser & Felton, 2007). Studies analyzing the poverty effects of commercialization of horticulture from an asset perspective are lacking.

Assets have been defined from different perspectives mainly by their importance based on the context of their respective study. Generally, assets can be classified into four broad categories:- physical capital, financial capital, human capital, and social capital (Attanasio & Szekely, 1999; McCulloch et al., 2001; Moser & Felton, 2007). These broad categories can be further split into various household indicators, for example, physical capital includes housing features such as roof type, floor materials and toilet type, as well as consumer durables such as radios, bicycles, television among others. Financial capital includes labour security (type of employment), productive durables such as sewing machines, transfer or rental income, bank savings and insurance policies

(Moser & Felton, 2007). Some of these asset indicators have been integrated into multidimensional poverty measurement approaches such as; the Human Development Index (HDI) (Anand & Sen 1994) and the multidimensional poverty index (MPI) (Alkire & Santos, 2010). Although easy to construct, these approaches have been challenged for their weaknesses in that they are based on predetermined sets of indicators and consequent weights, and are therefore subjective. To overcome the subjectivity problem, and maintain to a low-cost and reliable method of assessing the impacts of policies, programs and services, Zeller et al. (2006) developed a multi-dimensional poverty index using a principal component analysis (PCA), of a collection of household asset related indicators, broadly classified as human resources, housing assets, and productive assets. This method is adopted in this study to calculate a household asset index. However, unlike the index of Zeller et al. (2006) that use the value of assets, this study utilizes dummy variables to calculate the asset index.

3.3 Analytical framework and estimation procedure

A household can commercialize and earn income from vegetable production. Alternatively, it can produce other crops or engage in other farm or non-farm activities. The decision to participate in commercialization suggests a binary choice problem. In any given year, *t*, the choice of a household, *i*, between participating in commercialization or not will be determined by the expected utility associated with either option. If the latent variable $U_{i,t}^{comm}*$ represent the level of utility derived from commercialization, then:

$$U_{i,t}^{comm} = U_{i,t}(YCOMM_{i,t} + N_{i,t}, C_{i,t}, W_{i,t}) \tag{3.1}$$

where $YCOMM_{i,t}$ represents the income derived from the commercialization of horticulture, $N_{i,t}$ is family income from other sources (including labour, other crops and livestock enterprises, off-farm income, business and remittances), $C_{i,t}$ is the extra cost incurred from commercialization (for example transaction costs of meeting specific market requirements) and $W_{i,t}$ is a vector of farm and farmer characteristics that affect utility. The probability that a household participate in vegetable commercialization is therefore determined by a comparison of the expected utility of participation $U_{i,t}^{comm}*$, against the expected utility of not participating, say, $U_{i,t}^{nc}*$. In making this comparison, a household evaluates both the costs and benefits of participation. Such that the households will participate only if $U_{i,t}^{comm}* > U_{i,t}^{nc}*$. However, $U_{i,t}^{comm}*$ and $U_{i,t}^{nc}*$ are latent variables. We observe the actual participation of households in production of vegetables for different markets such that $U_{i,t}^{comm} = 1$ if

$U_{i,t}^{comm}* > U_{i,t}^{nc}*$ and $U_{i,t}^{comm} = 0$ if $U_{i,t}^{comm}* \leq U_{i,t}^{nc}*$. Participation in vegetable commercialization can therefore be represented as;

$$U_{i,t}^{comm} = Z_{it}\psi - \upsilon \tag{3.2}$$

where Z_{it} is a vector of covariates that affect commercialization, ψ is the vector of parameters, and υ is an error term with zero mean and variance, σ^2. Households are heterogeneous in characteristics such that some will participate in one market pathway, others in a different market, while others will not participate. It is however expected that, those who participate in vegetable commercialization in different market pathways; they will earn higher farm returns and household incomes, and thus positive poverty effects.

3.3.1 Estimation procedure

Household income is determined by household related factors (endogenous) and exogenous factors. Endogenous factors are related to household behaviour in making decisions on labour allocation to different activities- commercialization of vegetables or other crops, operating a business, seeking off-farm or on-farm employment and other household decisions. Thus, participation of households in vegetable commercialization may directly influence household income and thus poverty status. We postulate that the participation in vegetable commercialization has important positive marginal effect on household income and poverty, due to assured market access, higher prices of produce and better access to technology, including use of inputs. To model income and poverty effects, we build on a model used in impact assessment literature. We outline the origin of selection bias in carrying out impact assessment using panel data and the panel data techniques used to control for it. Consider a generic impact assessment specification:

$$Y_{it} = \mathbf{x}_{it}\boldsymbol{\beta} + Comm_{it}\,\gamma + M_i + \mu_{it}, \qquad i = 1, 2, .., . N;\ t = 0, 1; \tag{3.3}$$

where Y_{it} is the poverty outcome variable *(per adult equivalent income)* for i^{th} household at time *t* (*t=0 (2005) or t=1(2010)*.[12] A vector of observable explanatory variables is represented by $\mathbf{x}_{it}$, while $Comm_{it}$ is a vector of measures of vegetable commercialization (dummy or a continuous variable depending on the

12 The dependent variable can take any other poverty outcome other than per capita income such as per capita asset ownership. For this study, since we do not have the monetary values for assets, which could reflect the capital stock of a household, we limit the poverty effect analysis to per capita income.

specification) for household i during year t, M_i represents a vector of time-invariant unobservable variables and μ_{it} is the error term. The $\boldsymbol{\beta}$ and γ parameters are to be estimated.

Key explanatory variables ($\mathbf{x}_{it}$) that are likely to affect the poverty outcome include household demographic factors such as the gender of the household head, the dependency ratio, household size and the education level of household heads. For example, female-headed households are often disadvantaged with respect to economic well-being. A high dependency ratio is also associated with lower per capita income. Households with heads that are more educated are likely to have higher per adult equivalent income than less educated ones. Another set of explanatory variables that might influence poverty outcome are those related to the diversity of household income sources. For example, rural households that have access to off-farm income activities can use them to smooth fluctuations in crop income (Fafchamps et al., 1998). Other non-farm income sources such as businesses and remittances are important means of *ex-ante* diversification (Reardon & Berdegué, 2002). Land endowments (size and quality) are also included. Annual rainfall during the year prior to the survey and the covariance of rainfall variation during the year of the survey are important to capture weather shocks, risks, and other agro-ecological effects. Weather risk is observed as an important determinant of fluctuations in farm income (Hertel & Rosch, 2010), and is measured in this study as the coefficient of variation of monthly rainfall per household. Annual rainfall for the year prior to the year of survey provides a proxy for weather shocks that are likely to affect farm income position of the households in the consequent years. In addition, a dummy variable indicating whether households experienced any economic shocks between 2005 and 2010, such as drought or floods, loss of employment, loss of the major income earner, or other non-natural events such as civil conflicts. A drought for example can be a serious natural setback that leads to reduced assets, income, and consumption of rural households as observed by Dercon (2004) in rural Ethiopia. District dummy variables are also included to account for geographical location effects.

3.3.2 Estimation assuming that commercialization is exogenous

To start with, we assume the commercialization of horticulture is exogenous, that is, if the observed characteristics of respondents are controlled for, there are presumably no other factors that simultaneously affect commercialization and household poverty outcome. We estimate the model using participants and non-participants of each market pathway. In each formulation, the market pathway variables and a set of exogenous controls are included. Pooled ordinary least squares is then applied to estimate the model:

$$Y_{it} = \theta_t + \mathbf{x}_{it}\boldsymbol{\beta} + ExpoComm_{it}\,\gamma_1 + DomeComm_{it}\,\gamma_2 + Dist_i\xi + \upsilon_{it} \quad (3.4)$$

where $ExpoComm_{it}$ is the measure for commercialization through the export market pathway for household *i* at time *t*, and $DomeComm_{it}$ is the measure for commercialization through the domestic market pathway for household *i* at time *t*. The measure of commercialization is a dummy or continuous variable depending on specification. The parameter θ_t denotes a time-varying intercept, *Dist* is the district fixed effects, while υ_{it} is a composite error term that contains the measure of unobservable variables associated with the poverty outcome Y_i and is assumed in this specification to be uncorrelated with commercialization and with the exogenous control variables $\mathbf{x}_{it}$. The parameters γ_1 and γ_2 represents the effects of commercialization through export and domestic market pathways respectively on the poverty outcome, Y_i controlling for exogenous control variables $\mathbf{x}_{it}$. We hypothesize that the commercialization of horticulture through either of the two market pathways has positive effects and thus expect γ_1 and γ_2 to be positive and significant. However, the estimate of the γ_1 and γ_2 would generally be biased when commercialization is not exogenous. Participation in the commercialization of horticulture is likely to be non-random. For example, households with more resources or that possess better skills, capabilities and motivation can decide to participate and self-select themselves into the export market pathway, while those with fewer resources are more likely to engage in the domestic market pathway. Second, there might be geographical selection because farmers who are more distant to the market face high transaction costs of delivering their produce to the market as do traders who buy from such farmers. By ignoring such possibilities, the estimate of γ_1 and γ_2, may be biased.

3.3.3 Estimation controlling for self-selection to participation in commercialization

Assuming that commercialization is endogenous, the following estimation strategies can be used (i) standard fixed (within) regression estimator (ii) fixed effect model corrected for possible self-selection bias, and (iii) fixed effect instrumental variable approach. Each of these methods will be elaborated on and used in this analysis. In the presence of unobserved heterogeneity that is assumed to be correlated with one or more covariates, the fixed effects model is used. The panel nature of the data allows us to estimate a standard fixed effect regression model under the assumption of strict exogeneity of covariates $\mathbf{x}_{it}$ conditional on the unobserved effect (Wooldridge, 2002; 266). In the fixed effect model, the error term is expanded to include both time-invariant individual-specific effects M_i and a time-varying component μ_{it}. This differentiates the fixed effect equation from the naïve model (Equation (3.4) such that;

$$Y_{it} = \theta_t + \mathbf{x}_{it}\boldsymbol{\beta} + ExpoComm_{it}\gamma_1 + DomeComm_{it}\gamma_2 + M_i + \mu_{it} \qquad (3.5)$$

A standard household fixed effects model can provide consistent estimates of commercialization parameters under the assumption that all unobservable characteristics M_i, that influence the outcome are time invariant and removed by within transformation or first differencing (Wooldridge, 2002). However, there could be potential selection bias that may over- or under-estimate the causal relationships between participation in the commercialization of horticulture and per capita income. This is because individual specific unobservable characteristics may change over time, for example, an individual might improve commercialization skills as a result of continuous farming of vegetable particular crop or as a result of training in commercial farming - which happens often among farmers. This means that there could be a correlation between the commercialization through either market and the unobservable characteristics, contained within the time varying term μ_{it} resulting in bias. It is therefore important to test and account for selection bias.

Therefore, a more robust estimation that can account for selection bias, the Heckman framework, is used for the panel data. This involves estimating an equation for commercialization using a Probit model;

$$\Pr(Comm_{it,j} = 1 \mid \mathbf{Z}_{it}) = \phi(\mathbf{Z}_{it}\boldsymbol{\psi}_t) \ (j = Export\ market,\ Domestic\ market) \qquad (3.6)$$

where $\mathbf{Z}_{it}$ contains $\mathbf{x}_{it}$ and other variables that affects commercialization but not the poverty outcome Y_{it}. From Equation (3.6), we obtain an inverse mills ratio $(\hat{\lambda}_{it,j} \equiv \lambda(\mathbf{Z}_{it}\boldsymbol{\psi}_t))$ for each market pathway and for each year. The mills ratios are then included in Equation (3.5), such that;

$$\begin{aligned} Y_{it} = \theta_t + \mathbf{x}_{it}\boldsymbol{\beta} + ExpoComm_{it}\gamma_1 + DomeComm_{it}\gamma_2 + \\ \zeta_1 \hat{\lambda}_{it,ExporComm} + \zeta_2 \hat{\lambda}_{it,DomeComm} + M_i + \mu_{it} \end{aligned} \qquad (3.7)$$

While the two models described above (standard fixed effects and two-step fixed effect models) may address the problem of endogeneity due to unobserved heterogeneity, in the latter, the inverse mills ratios for the selection (commercialization) model are included to address the possibility of self-selection bias. Thus, this model provides consistent estimates of commercialization effects assuming a valid identification condition of the selection model variables. Identification involves inclusion of some variables hypothesized to be statistically associated with commercialization ($Comm_{it,j}$), but not with the outcome of interest (per adult equivalent income, Y_{it}). Transaction costs would generally affect the commercialization decision but not poverty outcome indicators. Thus, we use transaction cost variables to identify the commercialization decision. These are variables related to information access;

access to extension services, years of membership in a farmer's group, distance to the market and condition of the road and the price of important vegetable crops in the study area. Price variables used in the model are division mean values for French beans, snow peas, potatoes and cabbage deflated (2010 prices) or inflated (2005 prices) using a consumer price index (CPI) to a common month and year (February, 2009) and not specific household prices. [13]

In addition to unobserved heterogeneity and self-selection, there could be potential reverse causality of poverty outcomes affecting the decision to commercialize. If households with higher per adult equivalent income choose to commercialize through certain market pathways, this may lead to a simultaneity bias. To address this problem, household fixed-effects instrumental variable technique is employed. This technique involves estimating the commercialization Equation (3.6) simultaneously with the poverty outcome Equation (3.5). This can be expressed using the following equation:

$$Y_{it} = \theta_t + \mathbf{x}_{it}\boldsymbol{\beta} + (Comm_{it,j} = \boldsymbol{K}_{it}) + M_i + \mu_{it}$$
$$(j = Export\ market,\ Domestic\ market)\ t = 0,1\ i = 1,2,.,n \qquad (3.8)$$

where $\mathbf{x}_{it}$ represents a vector of overlapping variables that affect the poverty outcome Y_{it} and possibly commercialization, while $\boldsymbol{K}_{it}$ includes non-overlapping variables correlated with the extent of commercialization but not the poverty outcome. Equation (3.8) is estimated using fixed effects two-stage least-square estimator (FE2SLS). This estimation accounts for simultaneity bias as all variables are subjected to within transformation (Cornwell et al., 1992). To identify the commercialization equation, $\boldsymbol{K}_{it}$ will comprise transaction cost variables described above as those, that correlate with the extent of commercialization through a particular market pathway, but that do not directly influence poverty outcome.

3.4 Data and descriptive statistics

To measure the effects of commercialization of horticulture on household income and poverty, this study utilizes the two-wave survey panel data begun by the ICIPE in 2005–2006 and the follow up survey of the same households in 2011 by the author. Five districts were purposively selected from the two major vegetable producing provinces (namely Nyeri, Kirinyaga, and Murang'a of Central Province

13 French beans and Snow peas were identified as the important export market vegetable, while potatoes and cabbages as the most important domestic market vegetable crops.

and Meru and Makueni districts of Eastern Province) (Asfaw, 2008).[14] The Probability Proportional to Size (PPS) sampling method was utilized to randomly select 21 sub-locations within the five districts and 539 households for the survey interviews. A random sub-sample of 370 households was selected for the follow-up visits between July and August 2011 by the author. There was a sample attrition of 16%.[15]

The survey provided recall data, collected based on a structured questionnaire. The recall data provide information on household demographic characteristics, land use, agriculture production, household assets including livestock, agriculture related assets, dwelling types, off-farm income, and remittances and market access information including the type and characteristics of vegetables markets, access to credit, and farmer group membership.[16] Information on income and income sources allowed the calculation of income, while the types of assets are used to calculate the household asset index.

This study distinguishes the contribution of vegetable crops produced for export and those for domestic markets, treating production for each market as an independent farm enterprise unlike previous studies that have focused on either export or domestic markets. Therefore, in this study the sample of households is classified according to the commercialization market pathways - *export market pathway* (households who produce vegetables for the international market) and *domestic market pathway* (households who produce vegetables that are primarily consumed in the domestic market). The contribution of vegetables sold through the traditional markets have largely been ignored in past studies that places greater emphasis on export market vegetables (for example McCulloch & Ota, 2002) and those for modern domestic supply chains (domestic supermarkets) (for example Rao & Qaim, 2011). Further information about the classification of households into market pathways is provided in Chapter Two. Figure 3.1 shows the number of households under each market category. Households supplying jointly to both markets are classified as "*both export and domestic markets*".

---

14 Detailed information about the study area and data collection procedure is provided by Asfaw, (2008) and in Chapter 1 of this study.

15 To ensure that attrition bias does not exist in the subsequent models; T-test, Z-test and logistic regression (dependent variable dummy; 1=attrition and 0=non-attrition) was conducted using the variables given in Table 3.2. The results did not indicate significantly different variables between the attrition and non-attrition groups and thus n=309 (matched households in second round of the survey) is a representative sample based on observable household characteristics.

16 The recall period is 12 months (Jan-Dec 2005 for the first survey and Jan-Dec, 2010 for second survey or 12 months prior to date of survey).

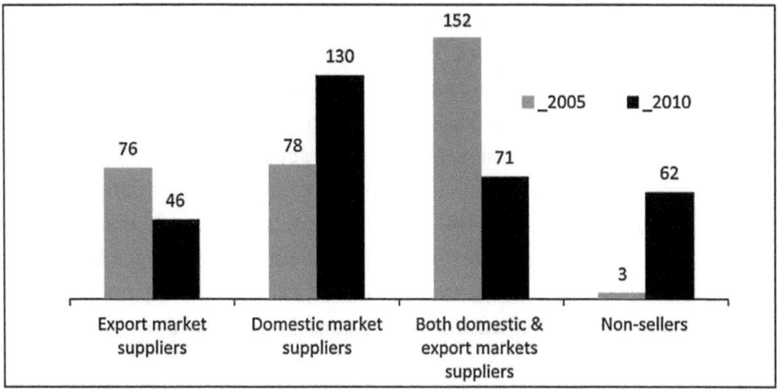

Fig. 3.1: Survey households by market pathway and survey year

3.4.1 Economic activities of smallholder vegetable producers

Descriptive analysis begins by examining the major and minor economic activities of sampled households and their dynamics over the two survey rounds. The results reflect alternative livelihoods of smallholders that had exited from commercial production of vegetables. Table 3.1 presents the major and minor economic activities of the household heads during the two survey rounds.

Table 3.1: Major and minor economic activities of the survey household heads

	Major activity			Minor activity		
	2005 (n=309) (%)	2010 (n=309) (%)	% Change	2005 (n=309) (%)	2010 (n=309) (%)	% Change
Horticultural production	72.5	41.1	–31.4	17.8	1.0	–16.8
Business	1.6	13.9	12.3	2.3	5.5	3.2
Cereal production	4.2	12.0	7.8	24.3	20.7	–3.6
Casual labour and other short–term work	2.3	6.8	4.5	5.8	2.6	–3.2
Civil servant	5.8	9.4	3.6	2.6	0.0	–2.6
Tea production	3.6	5.2	1.6	0.3	1.0	0.6
Coffee production	4.5	3.6	–1.0	1.3	1.6	0.3
Livestock production	3.9	3.6	–0.3	38.5	29.4	–9.1
Others	1.6	4.5	2.9	0.6	3.9	3.2
None	0.0	0.0	0.0	6.5	34.3	27.8

The reference as a "major" activity was made based on the activity that the household depends on most for its livelihood. About 31% of the households, that had cited vegetable enterprise as their major source of livelihood in 2005 did not consider this activity as their primarily means of livelihood in 2010. This shows the declining role of horticultural production among smallholders. A similar trend is demonstrated by farmers who considered horticultural production as their minor economic activity, while those who reported business as their main activity increased by about 12.3%. These businesses include cereals shops, consumer goods shops, farm input supply (agro-vets), food retail business (butchers and hotels), vegetable market brokerage among others. Possible reasons for the decline in commercial vegetable production are discussed in Chapter Two (section 2.4.2). Cereal (specifically maize) producers increased by about 8%. This was especially so in Nyeri district where farmers own relatively larger farms compared to the other districts included in the survey. The price for maize has remained high over the last two years in Kenya because of low precipitation levels, compelling farmers into commercial maize production. This not only enables them to benefit from the higher maize prices resulting from dwindling supply in the country, but also improves household food security. Other activities that exhibited positive change include casual labour and other short-term jobs. These are mainly employment in other farms, either producing vegetables or other farm activities such as tea and coffee production.

3.4.2 Demographic and socio-economic characteristics

Table 3.2 shows substantial differences in demographic and socio-economic characteristics between participants and non-participants of different market pathways. We performed tests for statistical significance of the differences in the means for these variables between households participating in each market pathway against non-participants of that particular market. The gender of the household head is predominately male across categories except between export market suppliers and non-suppliers in the year 2010. In 2010, export market suppliers had larger mean household sizes than non-exporters (5.63 and 5.13 respectively). Similarly, exporting households had a higher mean number of female adults (1.41) compared to non-exporters (1.23). This supports the argument that large households with large household size and working age females have higher labour resources and are thus more likely to engage in commercial vegetable production (von Braun et al., 1994). The dependency ratio for export market households is significantly higher than non-exporters in 2010.[17]

---

17 **Dependency ratio** is computed as the ratio of number of household members below 15 years and above 60 years to the number of household members aged between 15 and 60 years.

Table 3.2: Demographic and socio-economic characteristics of survey households

	2005		2010		2005		2010	
	Non-exporters	Exporters	Non-exporters	Exporters	Non-domestic sellers	Domestic sellers	Non-domestic sellers	Domestic sellers
Variable description	n=81	n=228	n=192	n=117	n=79	n=230	n=108	n=201
Head gender (1=Male)	0.91	0.91	0.87	0.95*	0.92	0.91	0.90	0.90
Head age (years)	50.21	45.3*	52.52	48.1*	43.70	47.5*	52.21	50.11
Dependency ratio	1.87	1.23	1.40	1.21*	2.00	1.19	1.37	1.30
Head education (years of schooling)	8.20	8.71	8.69	9.01	8.84	8.49	8.37	9.05
Total number of adults (15–60yrs)	3.80	3.32*	2.45	2.80*	3.41	3.47	2.32	2.73
Number of adult female (15–60yrs)	1.81	1.65	1.23	1.41*	1.67	1.70	1.20	1.35
Household size (adult equivalent)	4.71	4.41	5.13	5.63*	4.42	4.52	4.98	5.51*
Per adult equivalent income (*US$/year*)	501.9	593.3	643.4	679.2	542.3	578.6	616.7	678.5
Per adult equivalent income (less vegetable income, *US$/year*)	432.5	286.5*	546.2	423.6	273.9	342.3	493.2	503.3
Per adult equivalent income (*US$/day*)	1.37	1.63	1.76	1.86	1.49	1.59	1.69	1.86
Household annual income, *US$*)	1906.3	2299.3	2369.0	3475.2	2089.2	2233.0	2736.6	2815.4
Total household non-land asset (*index*)	3.557	3.559	3.746	3.826	3.568	3.555	3.587	3.878*
Per adult equivalent non-land asset (*index*)	1.001	0.9793	1.014	0.849*	0.965	0.992	1.028	0.9103
Off-farm employment (*dummy*)	0.5	0.4*	0.2	0.1*	0.4	0.5	0.1	0.2
Remittances (*dummy*)	0.2	0.2	0.4	0.3*	0.3	0.2*	0.4	0.4
Business (*dummy*)	0.2	0.2	0.4	0.4	0.1	0.2	0.4	0.4

*Note: *Significance at the 10% level or below (t-values (for continuous variable) and z-test (for categorical variables). The figures are means for continuous variables and proportions for categorical variables*

Table 3.3: Household income diversification by market pathways (mean %)

	Export market suppliers		Domestic market suppliers		Both export and domestic markets suppliers		Non-sellers		Total sample	
Income source	*_2005 (n=76)*	*_2010 (n=46)*	*_2005 (n=78)*	*_2010 (n=130)*	*_2005 (n=152)*	*_2010 (n=71)*	*_2005 (n=3)*	*_2010 (n=62)*	*_2005 (n=309)*	*_2010 (n=309)*
Vegetable income	47.33	28.71	24.04	26.58	48.17	44.33	0.00	0.00	41.41	25.64
Income from other crops	28.94	32.75	36.30	21.33	19.73	17.12	43.39	50.92	26.41	28.00
Off-farm income	9.32	3.11	19.36	11.65	14.59	5.79	46.24	9.85	14.80	8.67
Remittances	2.70	1.56	2.25	4.03	1.35	3.11	0.00	8.78	1.90	4.40
Business income	2.97	15.18	7.47	17.87	4.06	14.03	0.53	12.68	4.62	15.55
Livestock income	8.74	18.69	10.59	18.54	12.09	15.61	9.61	16.16	10.86	17.41

A simple comparison of mean per adult equivalent income between households commercially producing vegetables and non-sellers shows that the latter have lower income. This implies comparing sellers and non-sellers as it is done using cross-sectional data may provide biased results; hence, it is important to use panel data in analyzing the effects of market participation on income and poverty. To assess the importance of vegetables, we present per adult equivalent income calculated from total income and less income generated from vegetables. With this consideration, households supplying the export market in 2010 have significantly higher income compared to non-participants. This infers that vegetable income is important to the livelihoods of commercial horticultural smallholders.

3.4.3 Farm household livelihood diversification

Households in the study area hold diversified agricultural and non-agricultural related income portfolios. In addition to producing vegetables, households also practice dairy farming and produce basic food crops for local markets and home consumption such as maize, beans, rice and fruits and cash crops such as tea and coffee. A significant proportion of households are engaged in off-farm activities and business (Table 3.2). Income from agriculture constitutes about 80% of total household income. Table 3.3 shows the average shares of household income from different sources classified by vegetable market pathway.

Generally, vegetables, other crops and livestock account for the largest share of household income among the three market categories. The results indicate reduction of the role of vegetable commercialization. For instance, the share of vegetable income among households that specialized in export market vegetables declined from about 47% to 29% between 2005 and 2010. Similarly, the overall contribution of vegetables to household income decreased, by about 16% during this period. With respect to the total sample, business contribution to total household income increased by about 11% in 2010, while the share of income derived from livestock and livestock products increased by about 6%.

3.4.4 Asset ownership

The household asset index is constructed using a Principal Component Analysis (PCA) approach following Rutstein and Johnson (2004), Irungu (2002), Henry et al. (2003) and Zeller et al. (2006). Five key categories of assets are selected for developing the asset index, based on the asset portfolios reported during the two rounds of field survey. These asset categories were identified as most the important in the study area (Table 3.4). Construction of the household asset index

Table 3.4: Asset categories

Livestock assets (***L***),	Dwelling assets (**D**)	Consumer durables (***C***)	Agricultural assets (**A**)	Productive durables (***P***)
Cattle (calves, cows, heifers & bulls) Small ruminants (sheep, goats, rabbits, chicken) Traction/draft animals (oxen, donkeys & horses)	Iron roof, Permanent wall Piped water, Distance to water source(500m or less)	TV, Radio, Motor-bicycles, Bicycles	Hosepipes, Water pumps Sprinklers, Insecticide pump	Tractors Cars Ploughs Threshers Fridges Sewing machines Cart

is based on binary ownership indicators for each asset with exception of the livestock asset index, which was constructed using Tropical Livestock Unit (TLU).[18] Since we do not use the monetary value of assets, the asset index may not reflect the capital stock position held by the farmer rather it demonstrates ownership of assets in physical numbers.[19]

Table 3.5 below presents the changes in household non-land asset index values for the farmers commercializing through the export market, the domestic market, or through both markets as obtained from the PCA analysis.[20] Generally, a positive change in household asset index is observed across households participating in different market pathways. However, the total average household asset index value for non-sellers is lower in 2010 compared to 2005. Examining further the change in different assets among the non-sellers shows that, while agricultural and productive durables increased, the livestock, dwelling and consumer durables decreased. This implies that those who moved out of commercial vegetable farming had less livestock, as well as dwelling and consumer durables. The position of livestock assets among the non-sellers is also

18 The procedure of constructing the asset indices using the two methods is provided in Appendix 3A.

19 In absence of values of assets, the binary indictors of asset ownership and household characteristics can be used to assign the indictor weights using principal component analysis. This procedure is used in construction the DHS weath index (Rutstein & Johnson, 2004).

20 To make the indices comparable, once the factors are calculated using PCA analysis, they are then normalized to give them value between 0 and1. The formula for normalization is given as; Normalized value, $X_{Norm} = (X - X_{min})/(X_{max} - X_{min})$, where X is the initial value of the observation, X_{min} is the minimum observation and X_{max} is the maximum number of observations.

Table 3.5: Average non-land asset index values for surveyed households

	Export		Domestic		Both export and domestic markets		Non-sellers	
Asset index	_2005 (n=76)	_2010 (n=46)	_2005 (n=78)	_2010 (n=130)	_2005 (n=152)	_2010 (n=71)	_2005 (n=3)	_2010 (n=62)
Livestock	0.136	0.130	0.130	0.139	0.131	0.135	0.147	0.105
Dwelling assets	1.605	1.689	1.646	1.747	1.606	1.715	1.762	1.724
Consumer durables	0.501	0.518	0.486	0.541	0.464	0.538	0.572	0.480
Agricultural assets	0.996	1.079	0.948	1.136	1.033	1.152	0.753	0.846
Productive durables	0.330	0.325	0.347	0.312	0.321	0.342	0.308	0.319
Total	***3.569***	***3.740***	***3.557***	***3.876***	***3.555***	***3.882***	***3.541***	***3.473***

demonstrated in Chapter Two where the dynamics of smallholders' participation in the vegetable market is analyzed. Dwelling assets comprise the highest share of the household asset index values across the market pathways and survey rounds, while livestock assets took the lowest position. Increase in mean agricultural assets index values among the non-sellers could be an indication that those who switched from vegetable production, engaged in other farm enterprises that generated new or more assets.

Although land is considered in many studies as a major asset to farm households as well as in this study, it was not included in the asset indices. The reason for this can be explained contextually. One, unlike some assets (such as livestock) that may be relatively easy to accumulate, it was observed in the survey area that land ownership was more 'lumpy' and could take long time to acquire. Two, the research was undertaken in the Central and Eastern highlands of Kenya where population pressure on land is highest with an average household land ownership of 3 acres (1.2ha). Acquiring land in this area has further been impeded by the Government of Kenya through the Land Control Act, which was revised in 2010 with restrictions of the subdivision of agricultural land to parcels less than one hectare (GoK, 2010b: 24). As such, accumulation of land between the five-year panel period was considered unfeasible.

The majority of the households in this area are smallholders owning an average of 3.0 acres of land as shown in Table 3.6. Over 70% of the land is cultivated with an average of 2.2 acres across all households during the five-year survey period. Land fertility was observed to be a major constraint among the rural households, which is illustrated in the results for the small proportion of households that report owning fertile land (28%).

Table 3.6: Mean land ownership among surveyed household

Variable description	2005		2010		2005		2010		2005	
	Non-exporters (n=81)	Exporters (n=228)	Non-exporters (n=192)	Exporters (n=117)	Non-domestic sellers (n=79)	Domestic sellers (n=230)	Non-domestic sellers (n=108)	Domestic sellers (n=201)	Total sample (n=618)	
Land quality (*1=Fertile*)	0.32	0.24	0.28	0.34	0.22	0.25	0.29	0.30	0.26	0.30
Land cultivated (*acres*)	2.3	2.2	2.3	2.1	2.3	2.2	2.2	2.2	2.23	2.20
Land owned (*acres*)	3.7	3.0	3.5	2.4*	3.6	3.1*	2.9	3.2	3.20	3.08

*Note: *Significance at the 10% level or below. The figures are means for continuous variables and proportions for categorical variables.*

3.4.5 Poverty dynamics

In this study, two poverty outcomes are examined - *per adult equivalent income and per adult equivalent assets.* We examine the dynamics of poverty status of households using income poverty measure and relate income poverty trajectories to household asset accumulation quintiles. Using the income information presented in section 3.4.3, we constructed poverty trajectories that are classified as *consistently poor, ascenders, descenders and consistently well-off households*. Poverty trajectories are constructed by comparing per adult equivalent income (AE) per day for each year of survey (using US$1.25/day poverty line, adjusted for inflation to a common year, February 2009). For example, if Household **A** had AE/Income/day of $0.8 (<US$1.25) in 2005, and US$1.4 (>US$1.25) in 2010, then Household **A** is classified as "***ascender***" since it moved from below the poverty line**.** Similarly, households whose per adult equivalent income decreased from above to below the US$1.25 in 2010 are classified as "***descenders***". The "***consistently poor***" households are those who recorded per adult equivalent income below US$1.25 during two survey periods while "***consistently well-off***" stayed above the poverty line during the survey period. Figure 3.2 shows the distribution of the households participating in different markets by poverty trajectories in 2010.

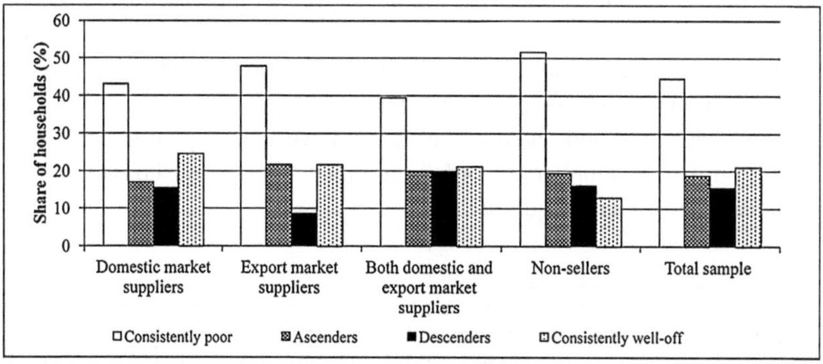

Fig. 3.2: Change in poverty trajectories among households participating in commercialization of vegetables for different markets

According to the government statistics, the central province (where three of the districts in this study are located) has the lowest incidence of poverty in the country (GoK, 2007b). The Kenya Integrated Households Budget Survey 2005/2006 conducted by the Kenya National Bureau of Statistics (KNBS) recorded 31% absolute poor (per adult equivalent) households in Nyeri district, 25% in Kirinyaga district and 29% in Muranga district (GoK, 2007b). In the Eastern province, poverty incidences differ across the districts due to large internal variability of agro-ecological zones and production systems. Meru district recorded 23% of households as absolute poor in 2005/2006, while the Makueni district recorded 64% (GoK, 2007b). Our survey showed a reduced incidence of poverty between the two survey years from an average of 64% to 60% of the total sample. However, our figures show considerably higher than average national rural poverty in 2005/2006 (36%). Nevertheless, we choose to categorize household by poverty trajectories as shown in Figure 3.2. The highest percentage of consistently poor households was recorded among non-sellers (52%), followed by export market producers (48%). Households supplying both export and domestic markets recorded the lowest percent of consistently poor households (39%). Those households who specialized in domestic market production in 2010 had the highest percentage of consistently well-off households (25%). Export market producers had the highest percentage of ascenders (22%) while producers of both domestic and export markets had the highest percentage of descenders (20%).

Based on the asset index developed in the previous section, assets accumulated during the two survey rounds were obtained by subtracting the asset index in 2005 from that of 2010. This difference was used to define asset quintiles as lowest, second, middle, fourth and highest. Figure 3.3 shows the distribution of households participating in different markets by asset quintiles in 2010. Similarly, to the information provided in Table 3.5, non-sellers recorded the highest percentage of households in the lowest asset accumulation quintile. Export market suppliers had the highest percent of household in the second lowest asset quintile while domestic and both market suppliers recorded the highest percent of households in the highest asset accumulation quintile.

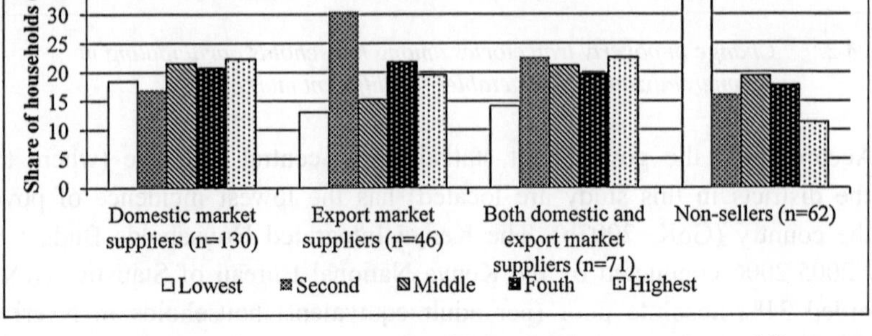

Fig. 3.3: Asset accumulation quintiles among household participating in different markets

As shown in Table 3.5, asset indices values increased between 2005 and 2010, implying that households may have allocated increasing income to accumulating assets (Table 3.2). Figure 3.4 presents the changes in the poverty trajectories by asset quintiles. Overall, positive asset accumulation across all the poverty pathways is estimated. Income descending household recorded the highest percent of households in the lowest asset accumulation quintile, while the consistently poor income households had the highest proportion of households in the highest asset quintile. The correlation between income and asset accumulation is high among the income ascenders, who reported the smallest share of households in the lowest asset accumulation quintile. Surprisingly, the consistently poor households had a bigger percent of households in the highest asset quintile compared to other poverty trajectories.

The above analysis reveals that horticultural commercialization may provide an important contribution to farm household income as non-sellers recorded a higher proportion of consistently poor than other groups. Similarly, with regard to asset, a bigger proportion of non-sellers are the category of the lowest asset accumulation quintile compared to the vegetable market participants. In the next section, the contribution of commercialization of vegetables to household income and poverty is tested empirically. Due the limitation of the asset measurement highlighted above, as dummy variables and not monetary values, the subsequent empirical analysis will only consider per adult equivalent income.

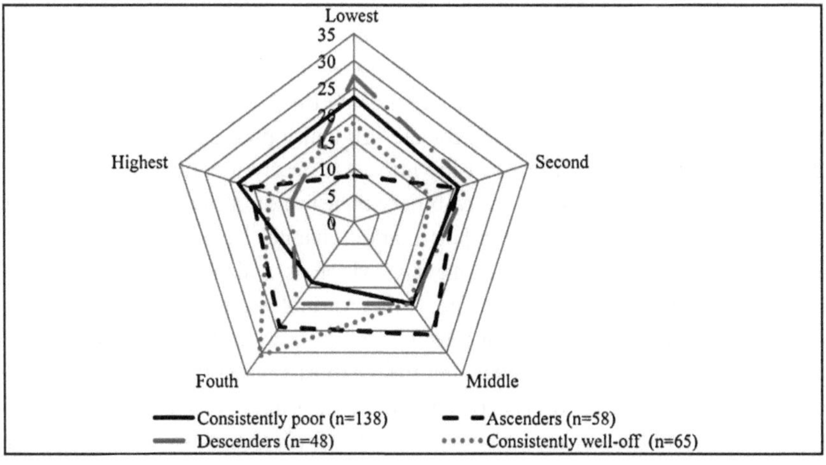

Fig. 3.4: Asset accumulation by poverty trajectories

3.5 Econometric analysis

The estimation results from the model specifications outlined in section 3.3 are presented in this section. To investigate the effects of commercialization on household income and poverty, different measures for commercialization each for both the export and domestic market are employed: a binary variable denoting participation and three horticultural commercialization indices (HCI) presented in section 2.3.3. Table 3.7 for example, displays the estimated coefficients for the binary variables denoting participation in the export (*export market*) and domestic (*domestic market*) markets in panel [1]. Similarly, panels [2], [3] and [4] presents the results for *HCI_1, HC1_2* and *HCI_3* respectively, all differentiated by type of the market. In order to test the poverty effects of vegetable commercialization among rural households, all specifications are separately estimated for the subsample of households whose income per adult equivalent (AE) and day was below poverty line in 2005 in addition to estimation on the full sample. Per AE income is used as the dependent variable representing the household poverty outcome. For the pooled OLS, standard fixed effect model and fixed effect instrumental variable models, only the most relevant results, commercialization parameters are presented. The results for the full results for these models are provided in the Tables 3B-1 in the Appendices.

Naïve OLS Estimation

The naïve pooled OLS estimates corresponding to Equation (3.4) are presented Table 3.7. The null hypothesis that all parameters are simultaneously equal to zero is rejected at 1% significance level for all the models. The results reveals that all measures of commercialization of vegetables through the export market pathway yield a positive and statistically significant effect on per AE income both for the total sample in column (1) and for the originally poor sub-sample in column (2). Interestingly, the magnitude of the impact is nearly identical in the total and reduced samples. With respect to the binary measure of participation, households that commercialize through the export market channel have approximately 40% and 42% higher income per AE and day, which is a sizable effect. Regarding the commercialization indices, holding other variables constant, per AE income for a household commercializing through export market increases by 0.5% for every 1%-increase in income generated from export vegetables out of total household income (*HCI_1_export*) or out of total crop sales (*HCI_2_export*), and by 0.7% for every 1%-increase in income generated from export vegetables out of the total annual value of vegetables produced (*HCI_3_export*).

Table 3.7: Pooled OLS estimates of the vegetable commercialization effects

		All households	*Poor households in 2005*
Models		*Per AE income (log) (1)*	*Per AE income (log) (2)*
[1]	Export market (dummy)	0.396*** (0.077)	0.421*** (0.089)
	Domestic market (dummy)	0.130* (0.080)	0.181** (0.093)
R-squared		0.386	0.31
[2]	HCI_1_export market	0.005*** (0.002)	0.005** (0.002)
	HCI_1_domestic market	0.0003 (0.002)	0.001 (0.002)
R-squared		0.368	0.2788
[3]	HCI_2_export market	0.005*** (0.001)	0.004** (0.002)
	HCI_2_domestic market	0.003* (0.001)	0.002 (0.002)
R-squared		0.373	0.2803
[4]	HCI_3_export market	0.007*** (0.001)	0.007*** (0.002)
	HCI_3_domestic market	0.005*** (0.001)	0.004*** (0.002)
R-squared		0.38	0.3024
Number of observations		618	392

Notes: Significant at 1% (***), 5% (**), 10% (*) probability levels; standard errors in parentheses; The sub-sample of households used in column (2) consist of households whose real income per AE and day in 2005 was below US$ 1.25 or households classified as consistently poor and those classified as ascending as described in Section 3c; AE is abbreviated for Adult Equivalent; Full model estimates are given in Tables 3B-1 in the Appendices.
*HCI_1=[Vegetable sales , hh,i; year/Total household income, hh,i; year]*100*
*HCI_2=[Vegetable sales, hh,i; year/Total crop sales, hh,i; per year]*100*
*HCI_3=[Vegetable sales, hh,i; year/Total value of vegetable crops produced, hh,i; per year]*100*

Commercialization through domestic market pathway also mostly shows a positive and statistically significant relationship with per AE income, except *HCI_1_domestic* in both samples and *HCI_2_domestic* in the sub-sample. Participation in domestic vegetable market is associated with per AE income that is 13% to 18% higher, all other things equal. Compared to commercialization through the export market, it is also with respect to the HCIs that the effects of domestic market commercialization are somewhat smaller, conditional on being statically significant. Considering the total sample, per AE income increases by 0.3% and 0.5% for every 1%-increase in income generated from vegetables supplied to the local markets out of total crop sales and from income generated from these vegetables out of the total value of vegetables produced respectively, holding other factors constant. The above analysis assumes that commercialization of vegetables is exogenous. As the decision to participate in commercialization is likely to be non-random, our impact analysis must account for potential bias due to unobserved heterogeneity and self-selection.

Standard Fixed effects Estimation

Given the panel nature of our data, Equation (3.3) can be estimated by the standard fixed effects approach. This is estimated as given in Equation (3.5) with the assumption of strict exogeneity of the control variables ($\mathbf{x}_{it}$). The standard fixed effects approach accounts for endogeity due to unobserved heterogeneity by expanding the error term is to include both time-invariant individual-specific effects and a time-varying component. The results of interest are shown in Table 3.8 while the full model results are given in Tables 3B-2. The F-test statistics indicate that not all parameters are jointly equal to zero at 1% significance level for all models.[21]

Similarly to the naïve OLS model, the standard fixed effects estimation results suggest a positive and statistically significant relationship between all measures of vegetable commercialization through the export market channel and per AE income both for the total and reduced sample. After controlling for potential endogeneity due to unobservable fixed effects, the coefficients are even slightly larger in magnitude compared to Table 3.7. Specifically, per adult AE for a

21 We tested the appropriateness of the fixed (within) effects model parameters against those of random effects (between) regression. While fixed effect (within) regression assumes strict exogeneity of explanatory variables conditional on the unobserved individual effects, random effects model on the other hand assumes that there is no correlation between the unobserved effects with the covariates (Wooldridge, 2002). The latter model treats household constant terms as randomly distributed across the entire household sample. By assuming that household specific error term is uncorrelated with other unobservable variables, random effects model faces a challenge. For example, in our study context, household characteristics such as ability and farming skills are likely to be correlated with the decision to participate in commercialization of horticulture and the level of commercialization. We ran Hausman test to determine if violation of this assumption should invalidate the use of random effect model. In all the models, the test statistics is significant ($p<0.05$), which is interpreted as evidence against the validity of random effects for all our models.

household supplying vegetables to the export market increases by about 0.65% for every 1%-increase in the share of income from vegetables out of total annual household income (*HCI_1_export*), by an average of 0.5% and 0.65% for every 1%-increase in the total share of vegetable income from the export market out of total crop sales (*HCI_2_export*), and out of total vegetables produced (*HCI_3_export*) respectively. The coefficients are slightly smaller for the household classified as poor in 2005 as presented in column (3). The estimates of the dummy variables however are of the same range as those of the OLS. Per AE income is approximately 41% and 38% higher for households supplying vegetables to the export market, for the total sample and poor households respectively, holding other variables constant.

Table 3.8: Household fixed effect (standard) estimates of the effects of vegetable commercialization

		All households	*Poor households 2005*
Models		*Per AE income (log)* *(1)*	*Per AE income (log)* *(2)*
[1]	Export market (dummy)	0.409*** (0.109)	0.381*** (0.134)
	Domestic market (dummy)	–0.088 (0.123)	0.0279 (0.153)
R-squared		0.20	0.24
[2]	HCI_1_export market	0.007*** (0.002)	0.006** (0.003)
	HCI_1_domestic market	–0.004 (0.003)	–0.004 (0.004)
R-squared		0.19	0.23
[3]	HCI_2_export market	0.006*** (0.002)	0.004** (0.002)
	HCI_2_domestic market	–0.001 (0.002)	–0.001 (0.003)
R-squared		0.20	0.23
[4]	HCI_3_export market	0.007*** (0.002)	0.006*** (0.002)
	HCI_3_domestic market	0.003* (0.002)	0.003 (0.002)
R-squared		0.20	0.24
Number of observations		618	392

Notes: (1) Significance at 1% (***), 5% (**), 10% (*) probability level; standard errors in bracket; Household fixed effects are jointly significant all the cases above; Full model estimates are given in Tables 3B-2 in the Appendices.

In contrast to the OLS estimates, commercialization through the domestic market exhibits no statistically significant relationship with per AE income except for *HCI_3_domestic*. Specifically, the share of income derived from domestic market vegetables out of the total value of vegetables produced yields positive and statistically significant coefficients on per AE income in the full sample (column 1). None of the commercialization specifications through those market exhibit significant coefficients among households classified as poor in 2005. The time-invariant individual-specific effects are removed by within transformation, which is conducted in standard fixed effect estimation above. However, as indicated in section 3.3.3, individual specific unobservable characteristics may change over time. This may lead to correlation between the commercialization and the unobservable characteristics contained within the time varying error term resulting to biased results. In that case, the two-stage estimation similar to the two-step Heckman framework applied on panel data as specified in Equation (3.7) is more robust than the standard fixed effects approach.

Correcting for selection Bias

Our main results in which we address the possible selection into commercialization and control for unobserved heterogeneity across households are presented in Table 9. The two stages estimation framework of panel data allows selection to be based on both individual-specific time-varying and individual-specific time-invariant unobservable characteristics. The model for per adult equivalent income is estimated in the second stage using fixed effects approach with covariates including the selection terms (inverse mills ratios) generated by estimating the selection Equation (3.6) for each market pathway and year of survey using Probit model.[22] The results for the full sample are presented in column (1) through (4), while those for households that were below the poverty line according to income per AE and day in the base year are presented in column (5) through (8). For the same dependent variables, the analysis start by investigating the binary indictor for commercialization first, column (1) and (5) and then the *HCI_1, HCI_2* and *HCI_3* in column (2) through (3) and in columns (6) through (8), respectively. For all the models, the *F*-test statistics indicate that not all the parameters are jointly equal to zero at 1% significance level. Household fixed effects are also jointly significant for all the cases.

22 The results for the probit model are not shown but available upon request. The estimation however is similar to that of Tables 2.4 and 2.5 in Chapter 2.

Table 3.9: Household fixed effects (selection-bias corrected) estimates commercialization effects

	All households				*Poor households 2005*			
	Per AE income (log)				*Per AE income (log)*			
	(1)	*(2)*	*(3)*	*(4)*	*(5)*	*(6)*	*(7)*	*(8)*
Export market (dummy)	0.416*** (0.114)				0.389*** (0.142)			
Domestic market (dummy)	–0.023 (0.128)				0.072 (0.163)			
HCI_1_export market		0.006*** (0.002)				0.006** (0.003)		
HCI_1_domestic market		–0.003 (0.003)				–0.004 (0.004)		
HCI_2_export market			0.006*** (0.002)				0.004* (0.002)	
HCI_2_domestic market			0.0003 (0.002)				–0.001 (0.003)	
HCI_3_export market				0.007*** (0.002)				0.006*** (0.002)
HCI_3_domestic market				0.004** (0.002)				0.003 (0.002)
Head education (years of schooling)	0.005 (0.023)	0.006 (0.023)	0.008 (0.023)	0.003 (0.023)	–0.009 (0.028)	–0.005 (0.028)	–0.005 (0.028)	–0.007 (0.028)
Household size (adult equivalent)	–0.120*** (0.026)	–0.11*** (0.026)	–0.118*** (0.026)	–0.120*** (0.025)	–0.08*** (0.031)	–0.08** (0.032)	–0.08*** (0.032)	–0.08*** (0.031)
Off–farm employment (dummy)	0.286* (0.121)	0.265** (0.123)	0.259** (0.120)	0.265** (0.119)	0.162 (0.156)	0.106 (0.157)	0.116 (0.154)	0.121 (0.152)

To be continued…

	All households				*Poor households 2005*			
	Per AE income (log)				*Per AE income (log)*			
	(1)	*(2)*	*(3)*	*(4)*	*(5)*	*(6)*	*(7)*	*(8)*
Business (dummy)	0.338*** (0.109)	0.338*** (0.112)	0.336*** (0.109)	0.329*** (0.109)	0.311** (0.140)	0.299** (0.145)	0.317** (0.142)	0.299** (0.140)
Remittances	0.084 (0.106)	0.047 (0.104)	0.044 (0.105)	0.072 (0.103)	0.349** (0.151)	0.298** (0.152)	0.298** (0.153)	0.328** (0.151)
Land cultivated (acres)	0.114** (0.054)	0.114** (0.054)	0.115** (0.054)	0.109** (0.054)	0.053 (0.078)	0.052 (0.082)	0.052 (0.082)	0.055 (0.082)
Annual rainfall (mm, lagged)	–0.002 (0.001)	–0.002 (0.001)	–0.002 (0.001)	–0.002 (0.001)	–0.002* (0.002)	–0.003* (0.002)	–0.003* (0.002)	–0.003* (0.002)
Covariance of variation (monthly precipitation)	–0.002 (0.011)	0.004 (0.011)	0.002 (0.011)	0.002 (0.011)	0.011 (0.014)	0.014 (0.014)	0.014 (0.014)	0.014 (0.014)
2010-dummy	–0.069 (0.218)	–0.103 (0.219)	–0.116 (0.217)	–0.065 (0.216)	0.018 (0.271)	–0.010 (0.274)	–0.025 (0.273)	0.009 (0.270)
Shock dummy	0.032 (0.131)	0.043 (0.132)	0.034 (0.132)	0.045 (0.131)	–0.059 (0.161)	–0.059 (0.160)	–0.065 (0.159)	–0.017 (0.157)
Mills ratio (export)	–0.027 (0.123)	–0.121 (0.118)	–0.085 (0.120)	–0.038 (0.121)	–0.018 (0.155)	–0.090 (0.151)	–0.071 (0.154)	–0.003 (0.156)
Mills ratio (domestic)	0.366* (0.208)	0.275 (0.202)	0.249 (0.204)	0.428** (0.206)	0.205 (0.269)	0.089 (0.254)	0.066 (0.257)	0.213 (0.260)
Constant	0.519 (1.669)	0.633 (1.662)	0.720 (1.671)	0.440 (1.649)	0.592 (1.876)	0.919 (1.856)	0.919 (1.928)	0.521 (1.899)
Number of observations	618	618	618	618	392	392	392	392
R squared	0.21	0.20	0.21	0.22	0.24	0.24	0.23	0.25
F(20,289) /F(20,176)	3.86***	3.66***	3.75***	4.00***	2.82***	2.74***	2.64***	2.88***

Note: *Significance at 0.01(***), 0.05(**), 0.1(*) probability level; standard errors in bracket. Household fixed effects are jointly significant all the cases above. The following control variables are also included in the above regression analysis but not shown in the tables since they are significant; head age (years) and its square, dependency ratio, remittances (dummy), total land owned (acres), head gender (1=male) and fertile land (dummy)*

The inverse Mills ratios for commercialization through the export market are not statistically significant in either specification, which implies there is no bias due to self-selection with regard to this market. In particular, in the absence of export market participation, there would be no significant difference on average between the current participants and non-participants of this market caused by unobserved effects. On the other hand, the null hypothesis that there is no significant difference between households who supply the domestic market and those who do not supply this market, is rejected in most columns as demonstrated by statistically significant inverse Mills ratios for the participation in the domestic market. Thus, the analysis of the impact of commercialization of vegetables through the domestic market must account for potential bias due to self-selection.

The results indicate that commercialization through export market pathway is positively and statistically significant related to per AE income, which is in line with the earlier results. The coefficients are again of similar magnitude as in Table 3.8. Specifically, per AE income for a household commercializing through export market increases by 0.6% to 0.7% for every 1%-increase in the share of income generated from export vegetables out of total household income (*HCI_1_export*), out of total crop sales (*HCI_2_export*), and out of the total value of vegetables produced (*HCI_3_export*). Similar results with slightly smaller coefficients are found in the reduced sample of (originally) poor households. The positive coefficient dummy that represents participation in the export market pathway also reflects similar results. Per AE income is approximately 40% higher for households commercializing through the export market holding other variables constant. After controlling for potential bias due to both time-invariant and time-varying unobservable household characteristics, participation in the domestic market does not appear to be strongly associated with per AE income. However, a positive and statistically significant association of 0.4%-increase in per AE income for every 1%-increase in the proportion of income generated from domestic market out of the total value of vegetables produced is found as presented in column (4) of Table 3.9.

Table 3.9 also presents the estimated coefficients of the included household observables that may systematically correlate with selection, as well as time-invariant and time-varying variables that are likely to influence per AE income. Similar covariates are used in the previous estimation. The two-stage fixed effects estimation shows that there is no significant influence on household per adult equivalent income by household human capital variables. The size of the household however has a negative and statistically significant relationship with per AE income. Off-farm employment, business ownership and size of cultivated land have positive and statistically significant influences on per AE income.

Households involved in business activities have the propensity to increase their per adult equivalent income by about 34%, while increase in size of cultivated land by 1 acre increases households' per capita income by about 11%. The positive and significant relationship between off-farm employment and business ownership emphasizes the positive role of diversification among smallholders. Contrary to expectation, the measures of weather shocks (*annual rainfall* for the year prior to the survey period) and weather risk (rainfall *coefficient of variation*) as well as the more general *Shock-* variable are statistically insignificant in all columns. However, pooled OLS estimates for these variables provide significant and expected signs (see Table 3B-1). High rainfall experienced in the previous period prior to the survey is positively and statistically significantly related to current periods' per adult equivalent income. This could be attributed to higher farm yields because of water availability. On the other hand, poor distribution of rainfall during the twelve months considered for survey, estimated in the model as the coefficient of variance, is negative and significantly related to per AE income (see Tables 3B-1). This emphasizes the role of weather change indicators in modeling farm household behavior, an aspect that has been ignored in previous studies that analyze the impact of commercialization of horticulture on farm household income (such as McCulloch & Ota, 2002; Omiti et al., 2007).

Fixed Effects Instrumental Variable Estimation

As mentioned in section 3.3.3, household fixed-effects instrumental variable technique is attempted due to potential reverse causality and simultaneity bias between per AE income and the decision and extent of commercialization. Endogeneity due to reverse causality may occur if households with higher per adult equivalent income choose to commercialize through certain market pathways. Fixed effects two-stage least-square estimator (FE2SLS) is used to estimate Equation (3.8). Again, the proxies for transaction costs identified as relevant in the study context are used as instruments for the export market pathway. These include access to extension services, number of years a household has been a member in a production and marketing farmer group, distance to the nearest market and condition of the road and division average prices for two important export vegetables (French beans and snow peas). The results of the parameters of interest from FE2SLS estimation are given in Table 3.10 (full model results are given in Appendix 3B-3). Due to the limitation of data, we fail to identify a good instrument(s) for commercialization through the domestic market. As such, we only present fixed effect IV model results for the export market pathway. The *F*-test statistics for all the model estimations indicate that not all parameters are jointly equal to zero at 1% significance level.

Table 3.10: Fixed effect instrumental variable effects of vegetable commercialization

		All households	*Poor households 2005*
		Per AE income (log) *(1)*	*Per AE income (log)* *(2)*
[1]	Export (binary)	0.790*	0.604
		(0.494)	(0.444)
R-squared		0.17	0.23
[2]	HCI_1_export	0.005	0.022
		(0.017)	(0.014)
R-squared		0.19	0.12
[3]	HCI_2_export	0.012	0.015
		(0.015)	(0.015)
R-squared		0.17	0.11
[4]	HCI_3_export	0.003	0.005
		(0.009)	(0.006)
R-squared		0.19	0.24
Number of observations		618	392

Note: Export market instruments: Export dummy =*Extension contact, distance to the nearest market, road condition, division level prices of French beans and snow peas ($/Kg); and years of group membership;* Extent of commercialization (HCIs) =*distance to the nearest market, division level prices of French beans and years of group membership; Significance at 1% (***), 5% (**), 10% (*) probability levels. Standard errors are given in parentheses.*

The results for the fixed effect instrumental variable approach for both the total and reduced samples are rather weak compared to those of the standard and self-selection corrected fixed effect analysis. All the coefficients are not statistically significant except the export dummy for the total sample (column (1), which shows that export market suppliers have per AE income approximately 79% higher than non-participants of this market. The transaction costs instruments are somewhat weak (the F-statistic for all the regression are below 5, see Table 3B-3), suggesting that the FE2SLS coefficients may be somewhat biased towards OLS estimates. Further, the exogeneity test of the instrumental variables across different measures of commercialization fail to reject the null hypothesis that commercialization through export market is exogenous (Table 3B-3). This again implies that, in our study, fixed effects models analyzed before would also yield consistent estimates, and thus the instrumental variable approach may not be required.

Overall, the results of the analysis find that returns from export market have significant effects on per adult equivalent income. Participation in this market is shown to alleviate income poverty through positive effects on per AE income in the total sample and in the sub-sample of poor households at base year. This is in line with past studies that use either cross-sectional or panel data to analyze poverty impacts of commercialization of horticulture (such as von Braun and Immink, 1994, McCulloch and Ota, 2002, Maertens et al., 2008, and Carletto et al. 2009). Even though the findings are not as robust, we find a positive relationship between participation in the domestic market and per AE income. Again, this agrees with previous studies as Rao and Quim (2012) who finds positive effect of participation of smallholders in vegetable market, although their study is an analysis of cross-sectional data on supermarket *vs* traditional market vegetable suppliers.

3.6 Conclusions and policy implications

This study descriptively analyzed the characteristics of smallholders participating in different market pathways and demonstrated how they have changed over time. The descriptive analysis have also demonstrated changes in two major poverty outcome- income and non-land asset-among households participating in different market pathways. Further, the study empirically analyzed the determinants of per adult equivalent income in light of vegetable commercialization. In the case of Kenya, this study is the first to analyze the marginal effect of commercialization of smallholder horticultural farming and thus shed some light to the ongoing debate of whether horticultural sector may be an engine for pro-poor growth in developing countries. Besides using panel data and thus captures market behavior of smallholder horticultural farms over time, this study contributes to the literature by investigating the effects of smallholder horticultural commercialization through different market pathways on households' income in Kenya. Descriptive statistics show that the share of vegetable income to total household income decreased between 2005 and 2010. Overall, there was a positive change in household income and asset accumulation between the two survey rounds. However, while on average per adult equivalent income increased, per adult equivalent asset decreased.

While starting from a naïve OLS estimation that assumes that the commercialization of vegetables to be exogenous, we move on to controlling for the unobserved heterogeneity across households, thereby adding to the existing literature that mainly relies on cross-sectional data. Further, we address the problem of potential selection into commercialization and employ a fixed effects instrumental variables approach in order to take account of possible reverse causality and simultaneity, and test all specifications in a sub-sample of households classified as poor in 2005. Our results are rather robust across specifications, thereby ameliorating our concerns about selection, reverse causality, and simultaneity: We find consistent evidence for a positive relationship between the commercialization of vegetables through the export market channel and income per adult equivalent. Considering the domestic market pathway, the naïve OLS specifications overestimates the effects of commercialization on household income and the results, furthermore, are less stable and suggest more strongly that selection into commercialization may be an issue.

This study suggests the need to measure poverty beyond household income or expenditure. Other livelihood aspects such as "household assets" should be considered to qualify commercialization of smallholder horticultural farming or agriculture in general as a rural "pro-poor" development strategy. In addition, the estimation of the effects of commercialization of agriculture can be improved by using panel data and addressing both heterogeneity and selection bias. Not controlling for such may overstate the impact of agriculture on household wellbeing. Pro-poor policies and livelihood diversification strategies should be enhanced. These include improving access to rural credit and encouraging savings among rural communities.

Further research should focus on intra-household distribution and utilization of income generated from the vegetable enterprises to provide a better understanding of the relationship between the commercialization of smallholders in horticulture and rural poverty. In addition, whereas this study has used physical counts of assets to develop the asset index, further research using monetary values of assets would be helpful on order to provide a better representation of the capital stock of a household.

Appendix 3A: Construction of the household asset index

a. Principle Component Analysis

Following Irungu (2002) and Zeller et al. (2006), I demonstrate how a multidimensional poverty index is generated using Principle component Analysis (PCA). In a PCA, a new set of variables (principle components) are created through linear combinations of the original variables. For example:

$$PC_i = X.V_i$$

where X is a $(n \; x \; k)$ matrix given data on k indicators for n households and V is a matrix of eigenvalues generated from the variance-covariance matrix of X. Vector V_1 contains the weights for each of the indicators X that are contained in the first principal component (Maddala, 1988, such that ;

$$PC_1 = X\begin{bmatrix} V_{11} \\ V_{21} \\ . \\ . \\ V_{k1} \end{bmatrix} \quad PC_2 = X\begin{bmatrix} V_{12} \\ V_{22} \\ . \\ . \\ V_{k2} \end{bmatrix} ....PC_k = X\begin{bmatrix} V_{1k} \\ V_{2k} \\ . \\ . \\ V_{kk} \end{bmatrix}$$

First principal component Second PC k^{th} PC

The first principal component accounts for the largest proportion of total variance in the set of indicators used, the second component accounts for the next largest amount of variance and so on for the higher order components (Zeller et al., 2006). These authors identified that the first component is always the one that should be considered as multi-dimensional poverty index since it accounts for the most of the movement (variance) of the indicators. They also carried out tests to justify the reason as to why the first principle the component is the most appropriate "poverty index." Having identified the first component as the 'poverty component,' one for each household (*j*), can be denoted as;

$$\text{Poverty index}_{\,j} = f_1 * \left(\frac{X_{j1} - X_1}{S_1}\right) + ... + f_N * \left(\frac{X_{jN} - X_N}{S_N}\right)$$

where f_1 is the weight for the first of the *N* poverty indicator variables that are identified as significant in the PCA, X_{j1} is the *jth* household's value (or dummy variable) for the first variable, and X_1 and S_1 are the mean and standard deviation of the first variable for all households.

b. Proportion to sample method

In order to check the robustness of the household asset index generated using the PCA, the asset index was recalculated using the proportion to sample method (following McCulloch & Ota, 2002; Asfaw, 2008). Interestingly, the methods arrive at a similar range as the household asset index. However, for further analysis in this study the figures generated using PCA are preferred as the method that has been used in most studies and that produces robust results.

This procedure involves use of the proportion of households with an asset as the weight for the indicator. The method provides a score for each asset owned by the household. The idea behind this procedure is that the costlier an asset, the wealthier a household needs to be to possess it (Rutstein & Johnson, 2004). The calculation of a composite household asset index involves a three-stage process.

First, it is necessary to construct a 1 by 1-column matrix (of 1's and 0's) that shows the asset ownership of each household. A household gets a value of 1 if it owns the asset and a 0 if otherwise. Each asset is represented by one column. *Second,* the cases of household with positive ownership (1's) for each asset category are summed up. The figure is then divided by the total number of households in the sample ($P_k = n_k/n$). This provides the probability of owning a certain asset, which is equal for the entire household sample. *Third*, the proportion of households with a particular asset is calculated using; $1 - P_k$. The proportion provides the weight of each asset. Thus, if a certain asset type is very common among households, the ownership of this asset adds a small score to the household asset index value, whereas the highest weights are assigned to the least possessed asset. (McCulloch & Ota, 2002; Asfaw, 2008). In an equation, the procedure can be described as:

$$\textit{Household asset Index}: A_i = \sum d_{ik}(1-P_k); \quad P_k = n_k/n, \quad \mathrm{i} = 1,\ldots.\mathrm{n}; \quad \mathrm{k} = 1,\ldots\mathrm{m}$$

Where d_{ik} =1 if a household owns the asset *k; P_k* is the probability of having asset *k; n_k* is the number of households which have asset *k,* n is the total household sample, and m is the number of household assets.

Appendix 3B: Full model estimates

Table 3B-1: Pooled OLS estimates of the effects of commercialization on household income

	All households				Poor households in 2005			
	Per AE income (log)				***Per AE income (log)***			
	(1)	(2)	(3)	(4)	(5)	(6)	(7)	(8)
Export market (dummy)	0.396***				0.421***			
	(0.077)				(0.089)			
Domestic market (dummy)	0.130*				0.181**			
	(0.080)				(0.093)			
HCI_1_export market		0.005***				0.005*		
		(0.002)				(0.002)		
HCI_1_domestic market		0.0003				0.001		
		(0.002)				(0.002)		
HCI_2_export market			0.005***				0.004*	
			(0.001)				(0.002)	
HCI_2_domestic market			0.003*				0.002	
			(0.001)				(0.002)	
HCI_3_export market				0.007***				0.007***
				(0.001)				(0.002)
HCI_3_domestic market				0.005***				0.004***
				(0.001)				(0.002)
Head gender (1=male)	–0.085	–0.055	–0.069	–0.091	0.165	0.195	0.188	0.160
	(0.125)	(0.127)	(0.126)	(0.126)	(0.156)	(0.159)	(0.159)	(0.157)
Head age (years)	0.027*	0.027	0.028*	0.026	–0.007	–0.011	–0.008	–0.008
	(0.019)	(0.019)	(0.019)	(0.019)	(0.021)	(0.021)	(0.021)	(0.021)
Head age (squared)	–0.0002	–0.0002	–0.0002	–0.0002	0.000	0.000	0.000	0.000
	(0.0002)	(0.000)	(0.000)	(0.000)	(0.000)	(0.000)	(0.000)	(0.000)
Dependency ratio	0.008	0.007	0.007	0.006	0.005	0.005	0.005	0.004
	(0.006)	(0.006)	(0.006)	(0.006)	(0.008)	(0.008)	(0.008)	(0.008)
Head education (years of schooling)	0.043***	0.042***	0.042***	0.043***	0.013	0.008	0.009	0.012
	(0.011)	(0.011)	(0.011)	(0.011)	(0.013)	(0.013)	(0.013)	(0.013)

	All households				Poor households in 2005			
	Per AE income (log)				***Per AE income (log)***			
	(1)	(2)	(3)	(4)	(5)	(6)	(7)	(8)
Household size (adult equivalent)	–0.171***	–0.166***	–0.169***	–0.172***	–0.107***	–0.104***	–0.106***	–0.109***
	(0.017)	(0.018)	(0.018)	(0.017)	(0.020)	(0.021)	(0.021)	(0.020)
Off-farm employment (dummy)	0.327***	0.337***	0.317***	0.296***	0.147	0.150	0.122	0.103
	(0.081)	(0.086)	(0.081)	(0.081)	(0.094)	(0.100)	(0.096)	(0.094)
Business (dummy)	0.474***	0.493***	0.466***	0.467***	0.430***	0.458***	0.426***	0.433***
	(0.079)	(0.083)	(0.080)	(0.080)	(0.092)	(0.099)	(0.095)	(0.093)
Remittances (dummy)	0.189**	0.179**	0.170**	0.184**	0.278***	0.273***	0.255***	0.278***
	(0.082)	(0.083)	(0.083)	(0.082)	(0.095)	(0.098)	(0.097)	(0.096)
Total land owned (acres)	0.012	0.011	0.008	0.003	0.014	0.011	0.008	0.008
	(0.019)	(0.019)	(0.019)	(0.019)	(0.020)	(0.020)	(0.020)	(0.020)
Land cultivated (acres)	0.141***	0.145***	0.154***	0.153***	0.084*	0.089**	0.096***	0.090*
	(0.032)	(0.032)	(0.032)	(0.032)	(0.036)	(0.037)	(0.037)	(0.036)
Fertile land (dummy)	0.209***	0.172**	0.175**	0.169**	0.176*	0.142	0.148	0.143
	(0.079)	(0.080)	(0.079)	(0.079)	(0.091)	(0.093)	(0.093)	(0.092)
Annual rainfall (mm, lagged)	0.002***	0.002***	0.002***	0.002***	0.000	0.000	0.000	0.000
	(0.001)	(0.001)	(0.001)	(0.001)	(0.001)	(0.001)	(0.001)	(0.001)
Covariance of variation (monthly precipitation)	–0.026***	–0.027***	–0.027***	–0.026***	–0.006	–0.005	–0.005	–0.006
	(0.006)	(0.006)	(0.006)	(0.006)	(0.008)	(0.008)	(0.008)	(0.008)
Year 2010 dummy	0.487***	0.444***	0.394***	0.524***	0.434***	0.386***	0.338*	0.464***
	(0.113)	(0.115)	(0.116)	(0.116)	(0.137)	(0.140)	(0.140)	(0.139)
Shock dummy	0.039	0.025	0.035	0.054	0.010	–0.013	–0.004	0.036
	(0.099)	(0.101)	(0.101)	(0.100)	(0.115)	(0.118)	(0.118)	(0.117)
Constant	–2.432***	–2.366**	–2.074**	–2.545***	–0.943	–0.802	–0.705	–1.111
	(0.940)	(0.955)	(0.960)	(0.945)	(1.166)	(1.191)	(1.192)	(1.173)
Districts fixed effects	Yes	Yes	Yes	Yes	Yes	Yes	Yes	Yes
Number of observations	618	618	618	618	392	392	392	392
F(22, 595)	17.0***	15.78***	16.13***	16.78***	7.6***	6.5**	6.5***	7.3***
R-squared	0.386	0.3684	0.3736	0.3828	0.311	0.279	0.280	0.302

Note: Significance at 0.01(***), 0.05(**), 0.1(*) probability level; Standard errors are given in brackets.

Table 3B-2: Household fixed effects (standard) estimates of the effects of commercialization on household income

	All households				Poor households in 2005			
	Per AE income (log)				Per AE income (log)			
	(1)	(2)	(3)	(4)	(5)	(6)	(7)	(8)
Export market_dummy	0.409***				0.381***			
	(0.109)				(0.131)			
Domestic market_dummy	–0.088				0.028			
	(0.123)				(0.139)			
HCI_1_export market		0.007***				0.006*		
		(0.002)				(0.004)		
HCI_1_domestic market		–0.004				–0.004		
		(0.003)				(0.003)		
HCI_2_export market			0.006***				0.004*	
			(0.002)				(0.003)	
HCI_2_domestic market			–0.001				–0.0008	
			(0.002)				(0.002)	
HCI_3_export market				0.007***				0.006***
				(0.002)				(0.002)
HCI_3_domestic market				0.003*				0.002
				(0.002)				(0.002)
Head gender (1=male)	–0.209	–0.128	–0.159	–0.219	0.056	0.163	0.141	0.052
	(0.260)	(0.261)	(0.259)	(0.260)	(0.282)	(0.276)	(0.278)	(0.283)
Head age (years)	0.026	0.038	0.036	0.026	0.008	0.013	0.010	0.005
	(0.039)	(0.040)	(0.040)	(0.040)	(0.033)	(0.034)	(0.034)	(0.033)
Head age (squared)	–0.0001	–0.0002	–0.0002	–0.0001	0.0001	0.0000	0.0001	0.0001
	(0.000)	(0.000)	(0.000)	(0.000)	(0.0003)	(0.0003)	(0.0003)	(0.0003)
Dependency ratio	0.010	0.011	0.011	0.010	0.008***	0.009***	0.008***	0.009***
	(0.008)	(0.008)	(0.008)	(0.008)	(0.002)	(0.002)	(0.002)	(0.002)
Head education (years of schooling)	0.007	0.007	0.010	0.006	–0.007	–0.005	–0.005	–0.006
	(0.023)	(0.023)	(0.023)	(0.023)	(0.032)	(0.031)	(0.032)	(0.032)
Household size (adult equivalent)	–0.128***	–0.118***	–0.12***	–0.130***	–0.083***	–0.076**	–0.080***	–0.084*

	All households				Poor households in 2005			
	Per AE income (log)				Per AE income (log)			
	(1)	(2)	(3)	(4)	(5)	(6)	(7)	(8)
	(0.025)	(0.025)	(0.025)	(0.025)	(0.031)	(0.030)	(0.031)	(0.033)
Off-farm employment (dummy)	0.243**	0.212*	0.218*	0.219*	0.132	0.076	0.095	0.097
	(0.117)	(0.119)	(0.116)	(0.116)	(0.141)	(0.136)	(0.137)	(0.141)
Business (dummy)	0.317***	0.308***	0.313**	0.305**	0.300***	0.285**	0.306***	0.291***
	(0.108)	(0.111)	(0.108)	(0.108)	(0.117)	(0.121)	(0.117)	(0.117)
Remittances (dummy)	0.082	0.028	0.033	0.064	0.339***	0.286**	0.289***	0.318***
	(0.116)	(0.116)	(0.115)	(0.116)	(0.115)	(0.111)	(0.115)	(0.113)
Total land owned (acres)	–0.010	–0.018	–0.017	–0.013	0.006	–0.003	–0.003	0.0005
	(0.028)	(0.028)	(0.028)	(0.028)	(0.035)	(0.036)	(0.037)	(0.036)
Land cultivated (acres)	0.109**	0.117**	0.115**	0.101*	0.047	0.057	0.056	0.0457
	(0.053)	(0.053)	(0.053)	(0.053)	(0.079)	(0.082)	(0.083)	(0.082)
Fertile land (dummy)	0.049	0.038	0.039	0.026	–0.033	–0.010	–0.020	–0.040
	(0.112)	(0.112)	(0.112)	(0.112)	(0.125)	(0.125)	(0.126)	(0.129)
Annual rainfall (mm, lagged)	–0.001	–0.002	–0.002	–0.002	–0.002*	–0.003*	–0.003*	–0.002*
	(0.001)	(0.001)	(0.001)	(0.001)	(0.001)	(0.001)	(0.001)	(0.001)
Covariance of variation	–0.001	0.007	0.005	0.003	0.012	0.018	0.017	0.014
	(0.011)	(0.011)	(0.010)	(0.010)	(0.013)	(0.013)	(0.013)	(0.013)
Year 2010 dummy	–0.001	–0.100	–0.087	0.021	0.075	–0.022	–0.033	0.083
	(0.200)	(0.195)	(0.193)	(0.197)	(0.195)	(0.180)	(0.199)	(0.202)
Shock dummy	0.006	0.001	–0.002	0.009	–0.082	–0.087	–0.087	–0.041
	(0.129)	(0.130)	(0.129)	(0.129)	(0.158)	(0.157)	(0.156)	(0.154)
Constant	0.584	0.274	0.4270	0.361	0.384	0.503	0.594	0.282
	(1.624)	(1.622)	(1.623)	(1.610)	(1.715)	(1.708)	(1.768)	(1.707)
Number of observations	618	618	618	618	392	392	392	392
Number of groups	309	309	309	309	196	196	196	196
R squared	0.20	0.19	0.20	0.20	0.24	0.24	0.23	0.24
F(18,291)/F(18,195)	4.07***	3.85***	4.04***	4.12***	4.36***	4.57***	4.24***	4.33***

Note: Dependent variable is Log income per adult equivalent. Significance at 0.01(***), 0.05(**), 0.1(*) probability level; Standard errors are given in brackets. Household fixed effects are jointly significant all the cases above.

Table 3B-3: Fixed effect instrumental variable effects of commercialization on per adult equivalent income

	All households								Poor households 2005							
	Per adult equivalent income (log)								Per adult equivalent income (log)							
	(1)		(2)		(3)		(4)		(5)		(6)		(7)		(8)	
Export market (dummy)	0.790**	(0.355)							0.604	(0.446)						
HCI_1_export market			0.005	(0.012)							0.022	(0.014)				
HCI_2_export market					0.012	(0.012)							0.015	(0.012)		
HCI_3_export market							0.003	(0.008)							0.005	(0.005)
Head gender (1=Male)	–0.271	(0.265)	–0.136	(0.257)	–0.187	(0.266)	–0.168	(0.282)	–0.020	(0.341)	0.153	(0.334)	0.126	(0.337)	0.047	(0.335)
Head age (years)	0.028	(0.039)	0.030	(0.044)	0.046	(0.047)	0.028	(0.042)	0.018	(0.051)	0.039	(0.059)	0.044	(0.065)	0.013	(0.052)
Head age (squared)	–0.000	(0.000)	–0.000	(0.000)	–0.000	(0.000)	–0.000	(0.000)	–0.000	(0.000)	–0.000	(0.001)	–0.000	(0.001)	0.000	(0.000)
Dependency ratio	0.011	(0.008)	0.010	(0.008)	0.012	(0.008)	0.010	(0.008)	0.008	(0.011)	0.012	(0.012)	0.011	(0.012)	0.008	(0.011)
Head education (years of schooling)	0.007	(0.023)	0.007	(0.023)	0.012	(0.023)	0.008	(0.023)	–0.007	(0.026)	–0.002	(0.028)	0.001	(0.029)	–0.003	(0.026)
Household size (adult equivalent)	–0.14***	(0.025)	–0.12***	(0.024)	–0.12***	(0.024)	–0.12***	(0.024)	–0.09***	(0.030)	–0.09***	(0.032)	–0.10***	(0.037)	–0.08***	(0.030)
Off–farm employment (dummy)	0.300**	(0.125)	0.225*	(0.134)	0.243*	(0.126)	0.207*	(0.120)	0.161	(0.155)	0.175	(0.164)	0.140	(0.161)	0.095	(0.143)
Business (dummy)	0.315***	(0.107)	0.334***	(0.108)	0.307***	(0.108)	0.311***	(0.109)	0.293**	(0.133)	0.352**	(0.145)	0.315**	(0.143)	0.286**	(0.134)
Remittances (dummy)	0.125	(0.120)	0.037	(0.113)	0.022	(0.116)	0.041	(0.112)	0.352**	(0.147)	0.252	(0.158)	0.227	(0.168)	0.303**	(0.143)
Total land owned (acres)	0.000	(0.029)	–0.016	(0.027)	–0.016	(0.028)	–0.015	(0.028)	0.009	(0.032)	–0.004	(0.033)	–0.003	(0.033)	–0.001	(0.031)
Land cultivated (acres)	0.097*	(0.053)	0.114***	(0.052)	0.119**	(0.053)	0.109**	(0.052)	0.047	(0.063)	0.074	(0.069)	0.067	(0.069)	0.052	(0.063)
Fertile land (dummy)	0.059	(0.111)	0.040	(0.109)	0.042	(0.110)	0.042	(0.109)	–0.022	(0.135)	0.023	(0.149)	0.027	(0.154)	–0.003	(0.140)
Annual rainfall (mm, lagged)	–0.001	(0.001)	–0.001	(0.001)	–0.002	(0.001)	–0.001	(0.001)	–0.002	(0.001)	–0.003*	(0.002)	–0.004*	(0.002)	–0.002	(0.002)
Covariance of variation (monthly precipitation)	–0.007	(0.012)	0.006	(0.010)	0.003	(0.011)	0.004	(0.012)	0.008	(0.014)	0.012	(0.013)	0.011	(0.014)	0.013	(0.013)
Year 2010 dummy	0.202	(0.257)	–0.110	(0.249)	–0.021	(0.252)	–0.087	(0.307)	0.157	(0.282)	0.182	(0.287)	0.061	(0.262)	0.048	(0.260)

	All households								Poor households 2005							
	Per adult equivalent income (log)								Per adult equivalent income (log)							
	(1)		(2)		(3)		(4)		(5)		(6)		(7)		(8)	
Shock dummy	0.005	(0.128)	0.009	(0.126)	−0.004	(0.128)	0.001	(0.129)	−0.084	(0.155)	−0.086	(0.165)	−0.090	(0.166)	−0.068	(0.155)
Number of observations	618		618		618		618		392		392		392		392	
F(17, 292)/ F(17, 179)	3.51***		3.32***		3.30***		3.34***		2.88***		2.58***		2.51***		2.86***	
R-Squared	0.17		0.19		0.17		0.19		0.23		0.12		0.11		0.24	
Exogeneity test (Davidson-MacKinnon test) F(1,291)	1.15		0.034		0.21		0.06		0.26		1.29		0.82		0.02	
P-value	0.28		0.85		0.65		0.81		0.61		0.26		0.37		0.88	
Underidentification test (Anderson canon. corr. LM statistic):	27.59		11.28		5.80		8.71		15.51		9.30		5.74		14.88	
Chi-sq(6) P-val	0.000		0.010		0.122		0.033		0.008		0.026		0.125		0.002	
Overidentification test of all instruments (Sargan statistic)	19.78		4.43		3.56		4.51		11.99		1.70		2.63		4.05	
Chi-sq(5) P-val	0.001		0.109		0.169		0.105		0.017		0.427		0.269		0.132	
Weak identification test (Cragg-Donald Wald F statistic)	4.69		3.66		1.85		2.80		3.01		2.94		1.78		4.85	
Stock-Yogo weak ID test critical values: 5% maximal IV relative bias	19.28		13.91		13.91		13.91		18.37		13.91		13.91		13.91	

Note: Export market instruments: Export dummy =*Extension contact, distance to the nearest market, road condition, division level prices of French beans and snow peas, and years of group membership;* Extent of commercialization (HCIs)=*distance to the nearest market, division level prices of French beans and years of group membership;* Significance at 0.01(***), 0.05(**), 0.1(*).

Chapter Four

4. Commercialization of Smallholder Horticultural Farming: Gender Roles and Implications for Household Well-being in Kenya

4.1 Introduction

The shift by smallholders from traditional markets to more innovative and modernized food supply chains entail important gender implications (Maertens & Swinnen, 2009). The commercialization of agriculture is often associated with new production technologies, marketing opportunities, and profit increases, which may reduce the role of women even if they were the main contributors of farm production before the shift. This is evident from the findings of previous studies such as for rice production in Gambia (von Braun et al., 1994), groundnuts in Zambia (Wold, 1997), French beans in Kenya (Dolan, 2001), and leafy vegetables in Uganda (Shiundu & Oniang'o, 2007). Focusing on vegetables, the rapid emergence and spread of high-value modern supply chains may have profound implications over the control of resources in rural households. Women are mostly excluded from contracting with large horticultural firms for the delivery of high-value produce (Dolan, 2001; Maertens & Swinnen 2009). For example, Dolan (2001), demonstrate how the expansion of the European vegetable market changed the domain of women in the control of vegetable production, resulting in reduced control over land resources, income, and labour contributions to household subsistence. Men appropriated land and labour resources traditionally used by women to cultivate vegetables for subsistence use and for sale in domestic markets to participate in the production of export vegetables (Dolan, 2001).

These modern supply chains are expected to have adverse effects and aggravate existing rural inequalities by excluding the poorest farmers, which are primarily women either by excluding them from the market or their exploitation by large agribusinesses (Farina & Reardon, 2000; Maertens & Swinnen, 2009). It had also been demonstrated that recent increases in international market regulations exclude rural poor smallholders (Okello et al., 2007: Muriithi et al., 2010a). Poor women farmers may be more affected due greater limitations on their capacity to comply with such regulations. Changes in gender roles due to commercialization may affect the bargaining power of women in intra-household resource allocation and thus directly or indirectly affect women and their families through the loss or diminishment of livelihoods. For example, Fischer and Qaim (2012a) find that that male control over banana output and revenues, a crop that prior to commercialization was primarily under the control of women , resulted in negative effects on household calorie consumption and dietary diversity. Quisumbing et al. (2006) observe that increases in the resources controlled by women can have greater development impacts than similar increases for men, as women are more likely to apply increased allocations towards children's education, health, and nutrition. Past studies on the gender implications of modern vegetable supply chains have paid more attention to the labour market (Barrientos et al., 2005; Tallontire et al., 2005; Maertens, 2009). Employment opportunities in this subsector in rural areas are, however, not only limited and thus employ a relatively small proportion of rural labour force but also make rural households vulnerable to rapidly changing global market variability. Thus, more attention is required to investigate gender dynamics that contribute to sustainable benefits from the product market.

In recognition of the role of women in the commercialization of farm enterprises and the consequent household welfare effects, as well as problems faced by women in production and marketing of their produce, several initiatives have been pursued by both the government and non-governmental organizations to link women and smallholders farmers in general to markets. Such initiatives include organizing farmers in production and marketing organizations (PMOs), improving access to market information, contract farming, and out-grower schemes (Njuki et al., 2011). Women are thought to have the most to win from these initiatives, for instance, PMOs are important for women farmers because they often have more limited access to productive resources than men (FAO et al., 2010). While the determinants of farmer participation in these initiatives (for example, farmer groups' membership, group sustainability), and their implication on household commercialization and welfare have been analyzed in past studies (Smartt & Haq, 2008; Barham & Chitemi, 2009; Markelova & Mwangi, 2010: Fischer & Qaim, 2012b), studies on the gender implications of such initiatives are very limited. A few cases include a study by Fischer and Qaim (2012a) that analyze gender implications of the commercialization of banana enterprises through collective action in Kenya. The study concludes that female group-membership have positive effects on female-controlled income share and consequently positive effects on household nutrition. The case study by Fischer and Qaim (2012a) is based on a domestic fruit supply chain, however such an analysis has not been performed in the framework of high-value vegetable production and particularly in the context of rapidly changing agricultural markets.

In addition to limited studies that explore the role of informal training, the role of human capital elements in agricultural commercialization has received little attention (Moti et al., 2009). Women remain significantly overrepresented among the illiterate (Tran-Nguyen & Zampeti, 2004). Improved access to education can be important for strengthening rural women's position in the labour process; in increasing women's capabilities through greater access to resources and information in order to represent themselves and for bargaining power in the allocation of intra-household resources Kirsten & Karaan, 2009: Moti et al., 2009).

From a policy perspective, the role of women's rural livelihoods is recognized. For example, the Comprehensive Africa Agriculture Development Programme (CAADP) policy focuses on small-scale farmers and women to improve agricultural productivity. In Kenya, the importance of gender in agriculture is recognized and incorporated in the Agricultural Sector Development Strategy (ASDS). This strategy proposes to develop a gender policy for the agricultural sector to achieve female empowerment and mainstream the needs of women, men and youth in all agricultural sub-sectors to ensure equal distribution of gains from various development initiatives. The policy further seeks to promote gender analyses and gender-based budgeting using participatory approaches at all levels of community-based development programmes (GoK, 2010a). The National Horticultural Policy recognizes the existence of gender inequality in the sharing of farming proceeds despite the fact that horticultural production is dominated by female farmers and also that gender discrimination exists in the horticultural labour market (GoK, 2012). The policy, however, highlights the limited understanding of gender issues as one of the major challenges to efforts to address them. This has also contributed to a lack of enforcement of gender-related labour laws in employment in the horticultural industry and lack of facilities to meet specific needs of women (GoK, 2012).

The insights on the gender implications of horticultural commercialization that will be generated from this study could therefore be important to the implementation of the National Horticultural Policy. Specifically, this study documents the different roles of women and men in the production of different vegetables, and in marketing and decision-making about the sharing of benefits. These roles could be interpreted as indicators of bargaining power of different household members. Second, the study evaluates the constraints and determinants that influence female participation in market-led agricultural initiatives such as horticulture. Female control over income and the management of plots allocated to commercial vegetable production are used as proxies for female participation. Focus is given to the roles of female group membership, female access to extension contact and training on vegetable production and marketing, market orientation of vegetables (domestic or export market) enterprises and other household and individual characteristics such as human capital elements. Finally, this study evaluates the impacts of female control over production and revenues generated from horticulture on household well-being. In estimating the impacts of female participation on household well-being, there is an endogeneity problem that needs to be mitigates through pursuit of different econometric strategies.

The rest of the chapter is structured as follows. In the next section, theoretical underpinnings on the linkages between gender, intra-household resource allocation and agricultural commercialization are provided. Subsequently, data and sample description is presented followed by detailed descriptive analysis on the role of gender in the smallholder vegetable production and marketing.[23] The empirical strategies and results are then provided before presenting conclusion and policy implications in the last section.

4.2 Theoretical underpinnings

Different studies have used different theories to explain intra-household allocation of resources. Using these methods, they examined whether the allocation of resources between male and female household members is efficient, and consequently household welfare outcomes. Most of the analyses of the impacts of horticultural commercialization are based on patriarchal perspectives that assume a unitary household model. This model is characterized by the aggregation of the preferences of all household members into a single joint utility. The unitary model assumes that within a household there is pooling of resources so that every household member can enjoy same level of welfare and that the household head is an altruist (Haddad et al., 1994). There is, however, qualitative and empirical evidence that contradicts the premises of this model (Chiappori, 1992).

The critics of the unitary model have provided an alternative theory, the collective model, which provides a more realistic representation of household behaviour (Alderman et al., 1995). In collective models, different household members may have distinct preferences such that the utility of each household member depends on their own and possibly that of other members (Chiappori, 1992). Collective model can be cooperative or non-cooperative. In non-cooperative models, household members operate as independent sub-economies with each member

23 An extensive and exhaustive descriptive analysis is necessary to supplement the empirical results for which we face difficulty in firmly addressing the problem of endogeneity.

controlling their income depending on their preferences. Conversely, in cooperative models, individuals have a choice either to remain individual or to pool all or some of their resources as a household or other sub-group (Chiappori, 1992).

A popular form of cooperative model is the Nash-cooperative bargaining model (Manser & Brown, 1980; McElroy & Horney, 1981). This model is based on the argument that a particular allocation (efficient resource allocation) is determined by the "threat point" of each household member. Under this model, a household is considered as a site for both cooperation and conflict where intra-household resource allocation is the outcome of a bargaining process (Agarwal, 1997). Chichilnisky (2008) describes gender gap as Nash equilibrium of a game with incomplete information about the work at home and in the market place. She concludes that a family produces a "local" public good (household welfare), using common properties or resources (labour hours of man and woman spent working in the home and in market place). Therefore, on the basis of these intra-household resource allocation arguments, defining measures such as poverty and other household level welfare indicators and assuming that all household members enjoy the same well-being outcomes can be misleading (Falkingham & Baschieri, 2009). The Nash-cooperative bargaining models could be related to past studies that have shown differential expenditure patterns from different farm enterprises between men and women. An example of early literature on this subject found that income from commercial crops are primarily controlled by men and used for non-food expenditures (E. Kennedy & Cogill, 1987). Recent studies include Njuki et al. (2011) who find that men spent only 6% of their income on food compared to 23% spent on the same item by women, and an empirical analysis by Fischer & Qaim (2012a).

Following the above theoretical understanding, in this study uses data from smallholders to analyze the different roles of men and women in vegetable commercialization including production and marketing decisions and the management of sales revenues. The study further focuses on the implications of the roles of women in the control over production and income on selected household welfare indictors. Differentiation of gender roles follows the household models of Carter and Katz (1992) and Katz (1995) that assume that allocation of resources between household members can be observed and measured. Using this model, Katz (1995) illustrated that complex processes that characterize domestic allocation of resources can be observed by examining how the resources (labour time, income, land, e.tc.), transfers, and expenditures (food and non-food) flow within a household.

The literature review and theoretical foundation discussed here leads to three hypotheses that address the research objectives. 1) The commercialization of vegetables contributes to increased male control over production (management of plot) and revenues. 2) Female oriented training and extension services on vegetable production and marketing, as well as membership in PMOs increase their control over production and revenue management and thus increase participation in the commercialization of horticulture and 3) Female control over vegetable production and revenues management has positive effect on household well-being.

4.3 Data

The methods employed in this study integration of quantitative and qualitative analyses applied to data collected in five districts within the Eastern (Meru and Makueni districts) and Central (Nyeri, Kirinyaga, and Murang'a districts) provinces of Kenya.[24] Extensive primary data collection was organized at different levels of supply chains: quantitative interviews with vegetable producers using household survey and qualitative interviews with agents of horticulture trading companies and focus group discussions with farmer organizations. The focus group discussions were conducted in two stages. The first stage was designed to examine farmer perceptions on change in gender roles related to vegetable commercialization s and the reasons they have come about in a joint discussion between men and women. The second stage involved single-gender discussions to elicit further information on gender roles in horticultural commercialization, the advancement of these roles over time, and their effects on different community groups.

The qualitative interviews covered five focus group discussions (or ten single-gender groups), one in each of the survey districts. Probability Proportional sampling was applied to guide the selection of households from each district for the quantitative household survey. A sample of 309 households was successfully interviewed. The data was collected between June and August 2011. The survey data provide details on household demographics characteristics, asset ownership differentiated by gender, crop enterprises, income and income sources (on-farm, and off-farm), sources of inputs, access to services (credit, extension services, training), marketing channels, and social capital indicators such as cooperative membership. Horticultural commercialization gender roles related to vegetable plot ownership and management, and income earning and management, were collected.

The information on the type of vegetables produced was used to classify households according to commercialization into *export market pathway* households and

24 Detailed information about the study area and data collection procedure is provided by Asfaw, (2008) and in Chapter 1 of this study.

domestic market pathway household.[25] An additional category was provided for households who supplied vegetables jointly to both markets. Owing to the nature of our analysis of comparison between men and women within households, the data utilized in this Chapter does not include women in female-headed household or men in male-headed household with no female spouse. Table 4.1 presents a summary of the sample household sample utilized for this analysis.

Table 4.1: Description of the survey sample by district

District of survey	Total Sample	Domestic market suppliers	Export market suppliers	Both domestic and export market suppliers	Non-sellers
Kirinyaga	76	24	12	21	19
Makueni	23	15	0	2	6
Meru	68	22	17	12	17
Murang'a	24	7	7	6	4
Nyeri	74	42	5	25	2
Total	***265***	***110***	***41***	***66***	***48***

Source: 2011 field survey data

4.4 Descriptive analysis

4.4.1 Household and farm characteristics

Table 4.2 shows selected household and farm characteristics of households considered in this analysis differentiated by market pathways. The profiles of households that did not produce vegetables for commercial purposes are also provided. There are some key differences across the market pathways. On average, heads and spouses in domestic market households have slightly higher ages of 50.2 and 43.4 years old, respectively, compared to 47.2 and 41.6 years respectively for export market households. Spouses from export market households have higher average education (9 years) compared to domestic market households (8.4 years). Domestic and export markets households have similar mean cultivated area. Export market suppliers however have smaller average areas allocated to vegetables (35.6 percent) compared to domestic market suppliers (44 percent).

Export market households have an insignificantly higher average per adult equivalent income per day (US$1.823), than the domestic market households

25 More information is provided in Chapters One and Two. In contrast to the previous chapters, in this chapter utilizes cross-sectional data from the second round of survey (2011)

Table 4.2: Selected household and farm characteristics of survey participants

	Domestic market participants (n=110)		Export market participants (n=41)		Both export and domestic market participants (n=66)		Non-sellers (n=48)	
	Mean	*SD*	Mean	*SD*	Mean	*SD*	Mean	*SD*
Per adult equivalent income (US$/day)	1.819	*4.612*	1.823	*2.218*	1.801	*3.572*	1.270	*1.535*
Per adult equivalent non-land asset index	0.928	*0.777*	0.749	*0.380*	0.747	*0.327*	0.773	*0.398*
Per adult equivalent food& clothing expenditure (US$ monthly)	18.7	*36.7*	12.7	*7.3*	14.2	*9.4*	14.9	*16.7*
Per adult equivalent expenditure on fees and health care (US$ monthly)	24.1	*120.8*	10.5	*14.6*	9.1	*9.9*	12.5	*24.7*
Female export revenue (%)			20.4	*36.9*	16.2	*32.8*		
Female domestic revenue (%)	18.6	*33.5*			24.1	*39.8*		
Female domestic plot manager (dummy)	0.3	*0.5*			0.4	*0.5*		
Female export plot manager (dummy)			0.4	*0.5*	0.3	*0.5*		
Remittances (1=Yes)	0.4	*0.5*	0.2	*0.4*	0.3	*0.4*	0.4	*0.5*
Head age (years)	50.2	*11.7*	47.2	*11.6*	47.9	*11.1*	54.3	*13.1*
Spouse age (years)	43.4	*12.2*	41.6	*10.4*	41.5	*9.4*	46.3	*11.9*
Household size (adult equivalent)	5.6	*2.3*	5.7	*2.0*	6.0	*2.3*	5.4	*2.0*
Dependency ratio	1.3	*0.8*	1.4	*0.6*	1.1	*0.5*	1.3	*0.9*
School going children	1.6	*1.3*	2.0	*1.2*	1.7	*1.2*	1.6	*1.4*
Children below 5 years	0.3	*0.5*	0.4	*0.5*	0.3	*0.6*	0.3	*0.5*
Female adults	1.3	*0.8*	1.4	*0.7*	1.6	*0.8*	1.4	*0.7*
Head education (years of schooling)	9.5	*3.7*	8.7	*2.4*	9.6	*2.9*	8.9	*3.5*
Spouse education (years of schooling)	8.4	*3.1*	9.0	*2.8*	9.0	*2.9*	7.7	*3.6*
Head major activity in 2010 (1=Non-farm)	0.4	*0.5*	0.3	*0.4*	0.3	*0.5*	0.3	*0.5*
Land owned (acres)	3.0	*3.2*	1.9	*2.0*	2.6	*2.3*	3.1	*3.0*
Land cultivated (acres)	2.2	*2.1*	1.8	*2.0*	2.3	*2.1*	2.6	*2.0*

	Domestic market participants (n=110)		Export market participants (n=41)		Both export and domestic market participants (n=66)		Non-sellers (n=48)	
	Mean	*SD*	Mean	*SD*	Mean	*SD*	Mean	*SD*
Vegetable area (% cultivated land)	44.0	*28.6*	35.6	*24.5*	60.2	*30.2*	7.6	*12.3*
Cash crop (1=yes)	0.3	*0.5*	0.5	*0.5*	0.4	*0.5*	0.3	*0.5*
Fertile land (1=yes)	0.3	*0.5*	0.4	*0.5*	0.3	*0.5*	0.2	*0.4*
Shock (*dummy*)	0.5	*0.5*	0.3	*0.5*	0.4	*0.5*	0.5	*0.5*
Distant to market town (km)	3.7	*3.8*	2.0	*1.5*	5.0	*6.3*	2.0	*2.7*
Road type (1=good)	0.7	*0.5*	0.5	*0.5*	0.8	*0.4*	0.7	*0.5*
Female member social groups (*dummy*)	0.90	*0.99*	0.51	*0.64*	0.64	*0.78*	0.75	*0.84*
Head major activity in 2005 (*1=Non-farm*)	0.2	*0.4*	0.0	*0.2*	0.1	*0.3*	0.1	*0.2*
Producer group member 2005 (*dummy*)	0.5	*0.5*	0.6	*0.5*	0.7	*0.5*	0.5	*0.5*
Female extension contact (*dummy*)	0.1	*0.3*	0.2	*0.4*	0.2	*0.4*	0.0	*0.0*
Female vegetable training (*dummy*)	0.1	*0.2*	0.2	*0.4*	0.2	*0.4*	0.0	*0.0*
Female non-land assets (%)	17.81	*10.10*	21.74	*14.75*	19.23	*14.78*	23.11	*13.44*
Jointly owned assets (%)	35.46	*32.77*	31.35	*32.49*	29.88	*29.49*	33.89	*34.91*
Annual rainfall (mm, 2009)	1007.7	102.8	1034.8	23.8	1041.1	85.2	982.6	129.1
Covariance of variation (*monthly precipitation, %*)	54.27	6.79	56.11	5.81	54.22	8.40	58.59	5.86
French beans price (US$/Kg)	0.64	0.39	0.56	0.26	0.49	0.19	0.46	0.18
Snow peas price (US$/Kg)	0.73	0.17	0.82	0.21	0.79	0.26	0.77	0.13
Potato price (US$/Kg)	0.25	0.06	0.27	0.05	0.25	0.06	0.26	0.04
Cabbage price (US $/Kg)	0.14	0.07	0.14	0.04	0.14	0.06	0.14	0.04
Maize area (% cultivated land)	23.16	17.50	19.69	18.16	18.66	15.76	30.20	21.62
Maize price (US$/Kg)	0.28	0.10	0.30	0.08	0.30	0.11	0.34	0.16

The exchange rate used is approximately Ksh.79 per US$. Vegetable prices are taken at division level.
Source: 2011 field survey data

(US$1.819). Turning to non-land assets, domestic market suppliers have higher average per adult equivalent non-land asset index values (0.928) compared to the export market participants (0.749).[26] Categorizing non-land assets by gender of the household member who owns the asset, females in export market households have higher average percent of assets out of total household assets (21.7%), than females in domestic market households (17.8%). Table 4.3 presents a breakdown of the household ownership represented by females among the survey households.

The information in Table 4.3 shows that women's ownership shares of most assets across all categories are less than 50% except for agricultural assets, of which they own reportedly owned over 40%. The higher reported female ownership of agricultural assets may reflect female participation in horticultural commercialization or simply agriculture in general. About 45% of the consumer assets are jointly owned by both male and female household members.

Table 4.3: Non-land household assets ownership

Asset category (n=265)	Type of assets	Percent owned by women (%)	Percent owned jointly (%)
Consumer assets	TV, radio, furniture, motorcycles, bicycle, cookers	3.36 [12.29]	43.99 [37.41]
Productive assets	Milking equipment, solar panels, batteries, store, sewing machines, refrigerators, car, cart, weighing machine, water tank	10.52 [18.89]	28.42 [43.35]
Agricultural assets	Spray pump, sprinkler, hose pipes, water pump, watering cans	41.05 [21.32]	23.97 [30.74]
Livestock assets	Calves, cow, heifers, oxen, bulls, chicken, sheep, goats, rabbit	13.39 [26.84]	35.18 [45.31]
Total assets		19.73 [12.86]	33.15 [32.25]

Note: Standard deviation are shown in brackets parentheses
Source: 2011 field survey data

Smallholder participation in the commercialization of horticulture is declining as observed in Chapter Two of this study. The trend may reflect negative effect on smallholder income and employment, particularly women, who are the prime producers and labourers suppliers for the vegetable markets. In light of this declining

26 The non-land asset indices were constructed using the Principal Component Analysis (PCA) approach following Irungu (2002), Henry et al. (2003) , Rutstein & Johnson (2004), and Zeller et al. (2006). The five key categories of assets given in Table 4.3 were identified as important in the study area.

trend, we extended the descriptive analysis to the households that had dropped out of commercial vegetable production to examine the activities of women in those households.[27] Deviating from the previous analysis (Table 4.2), this analysis includes female-headed households (the subsequent analyses however maintain the sample defined in Table 4.2). We differentiate vegetable market exitors into female-headed and male-headed households. The sample characteristics are given in Table 4.4. About 11% of the female-headed households that dropped out of commercial vegetable production depend on non-farm activities as their main livelihood source. An equal percentage of spouses from male-headed households that dropped out of vegetable production are involved in off-farm employment. A small percentage of spouses are involved in small rural businesses (18%), while a significantly large percentage of female-headed households receive remittances (67%). Similar to the results presented in Table 4.3, female members of the exiting households own more agricultural assets compared to the other categories of non-land assets (54%). Overall, female ownership of non-land assets in such households are about 25%. Many female-headed households that exited from commercial vegetable farming are also engaged in cash crop production (tea, coffee and rice) production (about 43%). Coffee is mainly found in Nyeri and a few areas of Kirinyaga, tea is found in Nyeri, Kirinyaga and Meru, while rice is produced in Kirinyaga.

Table 4.4: Characteristics of household that exited commercial vegetable production exitors

	Female-headed households (n=9)		Male-headed households (n=50)	
	Mean	SD	Mean	SD
Head education (years of schooling)	5.7	4.1	8.8	3.6
Age of household head (years)	60.7	13.8	54.1	12.9
Household head major activity (1=non-farm)	0.11	0.32	0.32	0.47
Household engaged in off-farm employment (dummy)	0.11	0.33	0.33	0.39
Spouse engaged in off-farm employment (dummy)			0.10	0.30
Household have business (dummy)	0.22	0.44	0.42	0.50
Spouse have business (dummy)			0.18	0.39
Household receive remittances (1=Yes)	0.67	0.50	0.40	0.49
Per adult equivalent income (US$/day)	1.15	1.41	1.24	1.52
Per adult equivalent asset (index)	2.40	1.66	0.81	0.44

27 Households described here are the ones who produced vegetables for sale in 2005 but not in 2010.

	Female-headed households (n=9)		Male-headed households (n=50)	
	Mean	SD	Mean	SD
Female consumer assets (%)	96.4	9.9	7.5	20.1
Female productive assets (%)	72.8	32.0	13.4	18.7
Female agricultural assets (%)	99.8	0.1	54.3	27.2
Female livestock assets (%)	86.0	20.1	19.1	31.2
Female total assets (%)	98.5	3.9	25.3	15.6
Household have cash crop (1=yes)	0.43	0.50	0.36	0.48
Total land owned (acres)	2.75	3.08	2.80	3.75
Land cultivated (acres)	2.09	1.51	2.27	2.52
Maize area (% cultivated land)	21.69	12.18	30.75	21.72

4.4.2 Gender roles in the commercialization of vegetable farming

Understanding how commercialization of small-scale farming activities affects the gender management of labour and land resources, income flows, expenditure patterns, food and nutritional security, and gender relationship is essential for addressing poverty in the rural communities (FAO et al., 2010). This section assesses the role of female and male members of the households in vegetable commercialization. The focus is on ownership and management of vegetable plots, membership in associations, and access to information through trainings or extension services on vegetable production and marketing and access to agricultural credit. Further examined in this section is the division of labour resources and control of income generated from vegetable enterprises.

4.4.2.1 Ownership and management of vegetable plots

Decisions on the utilization of land, including selling and leasing, are mainly taken by men across all districts surveyed and in a few cases negotiated between husbands and wives. This includes using land as collateral to obtain loans. In the Makueni district however, men were more likely to sell family land without consulting their spouses, underlining the differences in culture with regard to control of land resources between men and their wives. Across all the districts, men also dominated decision making about the type of commercial vegetable crops to be produced and the size of land to be allocated to each farm enterprise. In regions where export vegetables contribute a relatively higher share of household income, such as Kirinyaga district, men have greater influence on production of such vegetables while women have almost 100% control over production of vegetables for the domestic market. In Meru district, the vegetable crop selection decisions are made almost equally between men and women, and in some villages where there are other cash crops such as tea, these decisions are left to women. Figure 4.1 shows gender composition of intra-household ownership and management of vegetable plots by crop. Two major vegetables produced mainly for the export market and four types of vegetables for the domestic market are included in this analysis. On average, men own over 83% of the plots across both export and domestic markets vegetables. Men equally dominate the management of vegetable plots, although women have a substantial share, especially for domestic market vegetables. As demonstrated in previous studies, men typically manage vegetable production that has higher average returns with organized marketing channels. In this study, these include export vegetables as well as high-value domestic vegetables such as tomatoes and cabbages.

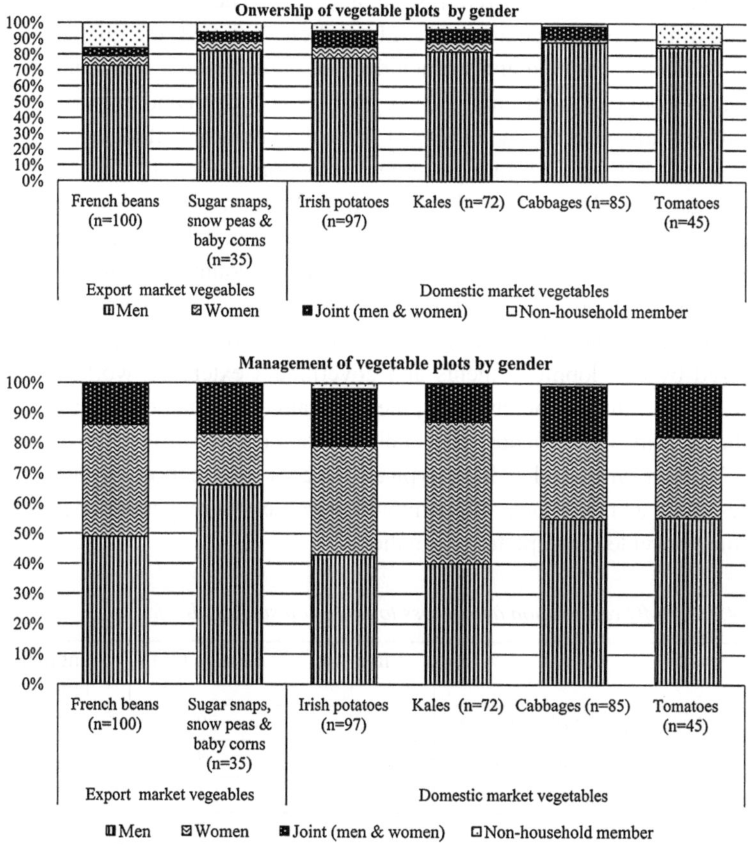

Fig. 4.1: Intra-household ownership and management of vegetable plots by gender (% of households)

The decision about which marketing channel to participate in, is made by the manager of the plot. In cases where a plot is jointly owned or/and managed, men consult their wives on the choice of market, but they have a stronger influence on the outcome. This is especially so where the household had no prior contact with the buyer contrary to the case of contract farming for export vegetable crops. Women freely make decisions about who to sell locally consumed vegetables that are under their responsibility.

4.4.2.2 Participation and registration in producer groups, access to training and extension services

Women in developing countries are often discriminated against in terms of access to agricultural information (Temu & Temu, 2005). The survey results show that household heads influence access to information. Although the qualitative information indicates that men and women decided equally about which member of a household would attend trainings on vegetable production and marketing, household survey showed more men (62% of those who reported attending such for the last 12 months) attended trainings compared to women (34%).

Men mainly rely on formal information sources, whereas women mostly rely on informal sources. This is also demonstrated in Table 4.5, which shows a higher percentage of men (63%) received extension contact in contrast to women (34%). Women reported a lack of time available to participate in various vegetable production and marketing trainings and other information forums, which are often organized by development agents and government extension workers. In addition to formal information channels, men exchange information informally while socializing with their friends, especially in social joints where they meet after work. Women on the other hand typically access information through informal sources, such as neighbors, and community women's groups and friends, which sometime provide incomplete information.

Table 4.5: PMO registration and access to extension services by gender

	Male (%)	Female (%)	Joint male and female (%)
PMO registered households (n=75)	68	29	3
PMO meetings attendance (n=75)	59	33	8
Vegetable trainings (n=101)	62	34	4
Extension contact (n=98)	63	34	3

Note: PMO abbreviated for Production and Marketing Horticultural Group
Source: 2011 Field survey data

Membership in producer groups is a common strategy for linking smallholder producers to the market. Results from this survey indicate that women are often excluded from attaining PMO membership. Over half of the women participants in the focus group discussions reported that they had to consult their husbands before joining a PMO. Unfortunately, in some cases consent is not always granted. However, in contrast to the requirement of landowners being registered with a farmer group (which was the case five years ago), PMO membership regulations have been modified to give men and women equal opportunities of joining producer groups.[28] This has not however been well received by men and has led to family conflicts. In Nyeri for example, men expressed their concerns that their wives were being "over-empowered" when they are registered in such groups as they receive their proceeds without involving their husbands. A similar observation was recently made by Miriti et al. (2011) in Meru district in the framework of a project that seek to promote high-value banana varieties through collective action. Introduction of the improved bananas, a crop that was mainly under the domain of women has positively influenced household well-being through increased high income, but has also brought intra-household conflict especially where wives are registered in those groups.

Few women (29%) are registered with PMOs relative to men (68%) (see Table 4.5). To ensure that registered female members are rewarded for their efforts, most of the PMOs require members to have a bank account to which proceeds are directed. This also provides women the opportunity to operate a bank account, a responsibility that many rural women had not previously embraced. Ownership of bank accounts among rural women can be interpreted as a process empowerment of women as it allows them to control the income generated from their farm enterprises. Although the decision to open a bank may sometimes be dictated by the PMO requirements or vegetable buyer, men have strong influence over who owns the bank account to which account the crop proceeds are paid. Attendance at PMO meetings is also dominated by men, which further indicates that women are deprived access to information and

---

28 In the early 2000s when PMO were reorganized to comply with GlobalGap standards, the registered PMO members had to be the owner of the farm, which is typically man and it was the same individual to whose bank account the proceeds were deposited. This arrangement denied women the opportunity to directly control income derived from high-value markets. Currently women are not required to present proof of legal ownership of land to register with a PMO, they can rent land for that purpose. This is an evolving concept which in some communities is highly contested, but that is embraced in others.

decision making in horticultural commercialization. Lack of market information is a major challenge that contributes to low participation of women in the commercial production of vegetables as a field manager from one of the export companies explained:

> *"Low participation of women in commercialization of vegetables is contributed by lack of market information. Although women are the actual farmers, men attend [..] meetings on their behalf. This implies that any information relied in such meetings, be it production, market information, etc., is given to the wrong audience who do not convey the same to the actual clients. The situation is more prevalent in some districts such as Muranga and Meru where there is high rate of women violence. Women are not allowed to attend meetings although they carry out most of farming activities. This situation is accelerated by abuse of alcohol by male household heads..."*
> (Interview with Sunripe Exporter field manager, Karatina, Nyeri district: 7/5/2012)[29]

4.4.2.3 Access to inputs and agricultural credit

Limited access to productive resources such as land presents a barrier to women for acquiring credit through the formal channels. The qualitative work revealed that men take the lead role in accessing agricultural inputs. Men typically determine the type and amount of fertilizers, pesticides and other farm inputs to be purchased, except in the few cases where women are more skilled in the utilization of those inputs. Men also typically control water for irrigation and credit access. Out of 37% of the households that reported to have received credit related to agricultural activities for the last 12 months, more men (54%) accessed credit compared to women (44%). However, over 80% of this credit was provided in the form of inputs by produce buyers and the cost deducted from sales. Women producers often seek the arrangements due the fact that they face significant constraints in acquiring financial resources to meet input costs. In the case of financial credit, men typically decide when and what collateral to use.

4.4.2.4 Division of labour in vegetable production

Vegetable production is labour intensive and requires extra labour during peak periods, especially during seasons when production activities for vegetables and other cash crops overlap. Past studies have observed that men usually decide when and who to hire during those peak labour requirement periods (Dolan, 2001). In so doing, they often allot more labour to their own plots (Udry, 1996).

29 Increasing alcohol consumption and abuse in the Central province in Kenya has been a public and government concern over the last few years. (Karugu, 2012).

In this study, it was observed that despite the fact that both men and women are capable of participating in any of the production and harvesting activities for vegetables, cultural constructs determine the gender division of labor. Men are usually responsible for manual tasks that require extra physical exertion, for example tilling. Men also typically take responsibility for irrigating crops as well as the application of fertilizers and pesticides. Men are also responsible for scouting crops, which involves monitoring vegetable crops to ensure they are not infested by pests and diseases. Some of these activities require knowledge and skills, which men gain from trainings and extension contacts whereas women in many cases do not get the opportunity to participate. Men are also in charge of seeking market information. Although women support their husbands in some of these tasks, they are also responsible for tasks that require deftness and attention such as planting, weeding, and harvesting. Transportation of produce to the collection centers is usually performed by women. During the peak seasons, especially when vegetables harvesting coincides with the harvest of other cash crops such as tea, households often need to hire labourers. Men also assist women in activities such as weeding during peak labour demand periods. Figure 4.2 shows the gender contributions to family labour inputs in the production and marketing of commercialized vegetables.

Generally, the results agree with previous studies that find that female household members spend more time on commercial vegetable crops than male household members. Significant differences in the number of labour hours between female and male household members are of observed in export vegetable-producing households (28 and 21 hours per week respectively). However, labour input sharing between male and female household members is evident among farmers specializing in domestic market production and those supplying both domestic and export markets. With respect to the total sample, there is an insignificant difference between female labour input in vegetable production and marketing (25 hours per week), compared to male labour input (22 hours per week). This finding contradicts many studies that conclude that women do most vegetables production work. However, most studies generalize about female labour input to both product and labour markets. From the labour market perspective, both for farm and agro-processors industry, women are often employed because of their dexterity in activities such as weeding, harvesting, and packaging. Moreover, women have limited labour opportunities compared to men and thus have high probability of working on large farms and agribusinesses. However, from the product market perspective, men and women share responsibilities, mainly based on cultural constructs. This is especially the case in households for which where vegetable enterprises are major source of livelihood.

Decisions on labour allocation are dominated by men. Across all of the districts, qualitative work revealed that women that operated independent plots were required to work in their spouses' plots before engaging on their own plots. Sadly, men did not provide reciprocal assistance to women in their plots. Men also decided whether extra labour should be hired, but in consultation with their wives.

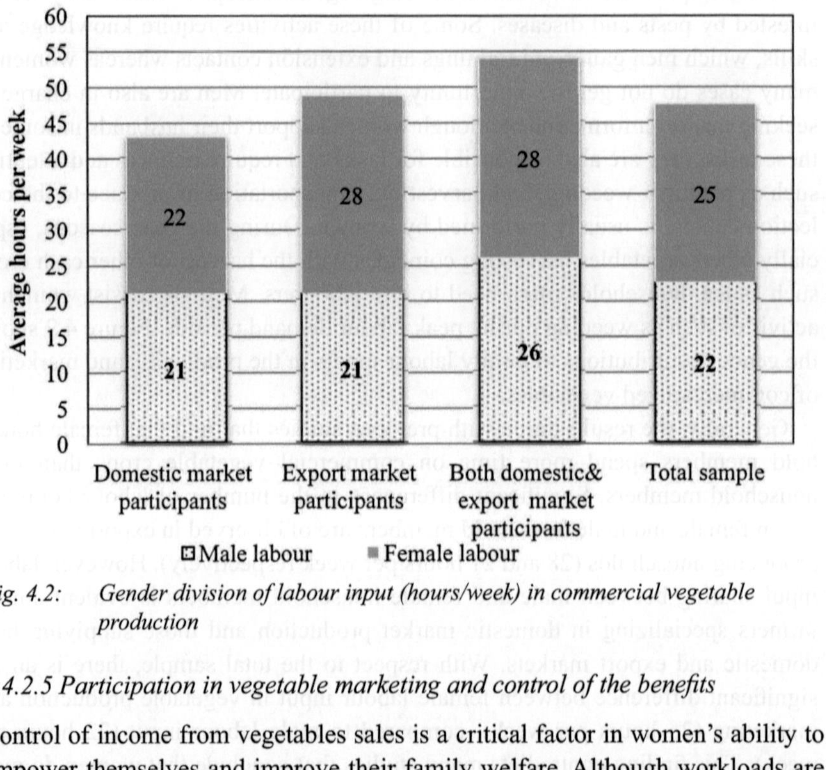

Fig. 4.2: Gender division of labour input (hours/week) in commercial vegetable production

4.4.2.5 Participation in vegetable marketing and control of the benefits

Control of income from vegetables sales is a critical factor in women's ability to empower themselves and improve their family welfare. Although workloads are shared between men and women, women have limited access and control over the benefits. This reflects gender inequality in the production of vegetables and benefit sharing, as we will illustrate below. The nature of market orientation differs between women and men. Access to information through formal sources provides men with the advantage of being able to negotiate higher prices for their produce because they have a better bargaining position than women based on acquired market information. Women tend to accept the prices offered on site since they do not have full knowledge of prevailing market prices. In some cases, men travel to secure higher prices for their produce or contact traders from distance places who offer them better prices, while women prefer selling to traders and consumers whom they can reach on foot.

There are significant differences in the proportion of households in which women received money from sales across different types of vegetables (Figure 4.3). Generally, men received most income from export market vegetables. In 35% of the households that produced French beans for example, women exclusively received money from sales, while males received the money in 59% of the households. In the remaining 6%, the money from French bean sales was received either by a male or by a female household member. A similar observation is made with the domestic market vegetables that fetch higher returns such as Irish potatoes, cabbages and tomatoes. Conversely, the figures for kales production show that the percentage of households in which women exclusively received money from sales increased (71%) compared to men (25%). This was also reflected during focus group discussions in which women noted that kales were the most preferred vegetable in terms of receipt and control of income. Kale however, provides the least returns compared to other types of vegetable. The results support other previous research findings that women are likely to receive and control income from commodities that generate lower average revenues, typically dominate commodities with higher revenues and that are often sold in formal markets (Njuki et al., 2011).

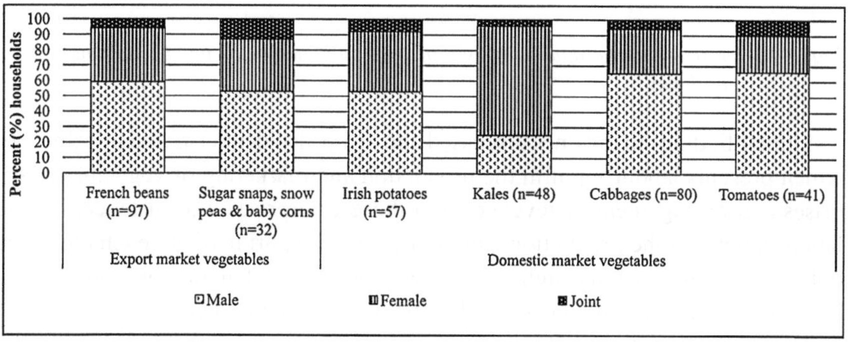

Fig. 4.3: Income recipient by gender for different vegetable crops (% of households)

In terms of the management of vegetable income, men autonomously manage a bigger share of income from both export market vegetables (26%) and domestic market vegetables (28%) (Figure 4.4). A similar observation was echoed across all the survey districts during the focus group discussions. It was also observed that in most cases men do not disclose to their wives the amount of income they receive from the sale of vegetables. In contrast, men typically demand that their wives should reveal their earnings and decide together how it should be spent. As a result, women sometimes opt to sell smaller quantities in secret, for example selling part of their export vegetable crops to brokers, while the household has a contractual arrangement with an exporter. This is often referred to as side selling, which threatens the sustainability of existing contractual relationships.

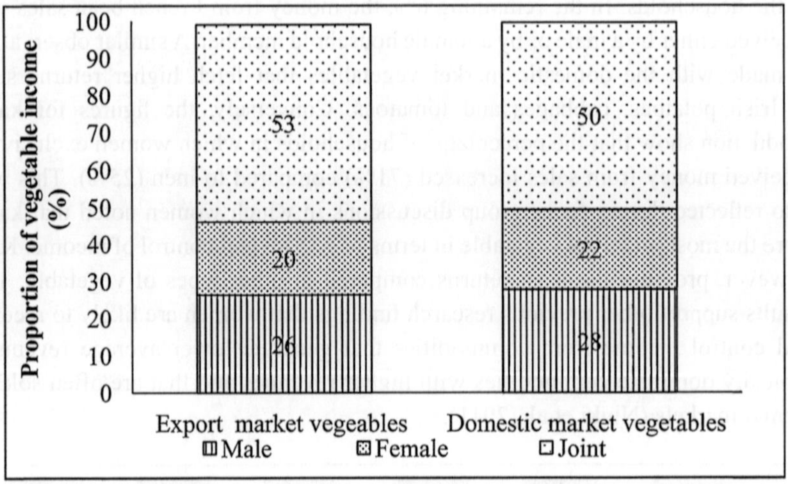

Fig. 4.4: Intra-household management of vegetable income by gender and market pathway

Similar findings to the ones discussed above are also demonstrated in Table 4.6, which presents labour input in the production and marketing of vegetable enterprises and management of revenue from those enterprises. For instance, female labour input into the production and marketing of export market vegetables does not result in significantly higher revenues managed by female household members. In contrast, the higher the male labour input, the higher the proportion of income from export market vegetables managed by a male household member ($p<0.05$). For domestic market vegetable producers, higher female labour input translates into a significantly higher proportion of income managed by a female member ($p<0.05$). In the case of households that supply both export and domestic markets, female labour input does not significantly relate to the proportion of income managed by female members of the household, however, the opposite applies to male labour input.

Table 4.6: Correlation between family labour input and revenue control in commercialized vegetable enterprises by gender

	Female labour	*Male labour*
Export market vegetable suppliers	0.034 (0.834)	0.310** (0.052)
Domestic market vegetable suppliers	0.187** (0.051)	0.284*** (0.003)
Both domestic and export market suppliers		
Export market vegetable sales	0.029 (0.819)	0.382*** (0.002)
Domestic market vegetable sales	0.041 (0.743)	0.2967** (0.016)

Note: P-values of each correlation coefficient are given in parenthesis. Significance at 0.01(***), 0.05(**), and 0.1(*) probability level.
Source: 2011 field survey data

Income indicated as managed jointly by men and women is mainly spent on common household expenditures such as school fees, house improvements and other household investments. The share of income managed by women is mainly spent on food and children requirements such as clothes. It was observed that a significant number of men used their share of income to purchase adult goods such as beer and other expenditures that do not benefit the family. Women's responsibility for household food security and childcare was also highlighted during the focus group discussions in which women ranked food and clothing as their first and second expenditure priorities for income received from vegetable sales. Men on the other hand ranked children school fees and investment in land as their first and second priorities.

In addition to the information provided in Figure 4.3, Figure 4.4, and Table 4.6, further analysis is done to assess whether high levels of commercialization result in increased male control over production (management of vegetable plots) and control of over revenues. This is done by utilizing the measurement of commercialization described in Chapter Two, specifically commercialization as the percentage of income derived from sale of vegetables out of total household income (Horticultural Commercialization Index (HCI_1). Using this index, households producing and selling vegetables are classified into two groups. If HCI_1 is equal or greater than the median HCI_1, the household is classified as "high vegetable commercialization" while those with HCI_1 below the median HCI_1 are classified as "low vegetable commercialization". This is calculated for each market pathway. The results of the descriptive overview of gendered control over income generated from vegetables sales and the management of vegetable plots are given in Table 4.7 and Table 4.8 respectively.

Table 4.7: Intra-household control over vegetable revenues by gender, market pathway and level of commercialization

	High commercialization	*Low commercialization*	*Difference*
	Share of vegetable income (% mean)	Share of vegetable income (% mean)	t-test[a]
Export market vegetables	n=53	n=54	
Male	36.7 (45.6)	16.5 (32.1)	20.2***
Female	14.3 (31.19)	21.2 (37.1)	–6.82
Joint	49.0 (48.1)	59.2 (46.0)	–10.15
Domestic market vegetables	*n=88*	*n=88*	
Male	37.9 (42.4)	17.1 (35.6)	20.8***
Female	13.4 (28.0)	28.0 (41.4)	–14.6***
Joint	47.4 (45.0)	52.1 (46.7)	–4.7

Note: Standard deviations are provided in parentheses; [a]two-sample t-test with unequal variance. Significance at the 0.01(***), 0.05(**), and 0.1(*) probability levels.
Source: 2011 field survey data.

Table 4.7 shows the distribution of the management of vegetable income across the households, disaggregated by groups of households with low or high commercialization levels. Male control is significantly higher under high level of commercialization for both export and domestic market vegetables, suggesting that male control increases with the level of commercialization. The percentage of female control increases more than that of men under low commercialization levels again in both export and domestic markets. A similar scenario is observed when we consider the proportion of male managers of vegetable plots was considered (Table 4.8). French beans and Irish potatoes, the two most important export and domestic vegetables respectively, present a significant proportion of male managers compared to females who manage similar plots.

The results in Table 4.7 and Table 4.8 show that the commercialization of vegetable production may contribute to greater male control over vegetable production and the management of resulting proceeds. However this analysis is based on correlation and does not provide evidence of causal relationships or control for any other factors

Table 4.8: Intra-household management of vegetable plots by gender, market pathway and level of vegetable commercialization

		High commercialization	*Low commercialization*	*Difference*
		Proportion of vegetable plot managers (%)	Proportion of vegetable plot managers (%)	Pearson Chi2 (1) [a]
Export market vegetables				
French beans	Male	73.3	26.1	20.1**
	Female	17.8	58.7	
Sugar snaps, snow peas and baby corns	Male	71.4	55.7	0.1
	Female	21.4	22.2	
Domestic market vegetables				
Irish potatoes	Male	69.8	31.8	8.6***
	Female	30.2	68.2	
Kales	Male	53.3	47.4	0.2
	Female	46.7	52.6	
Cabbage	Male	72.7	56.5	1.8
	Female	27.3	43.5	
Tomatoes	Male	92.9	52.2	6.6***
	Female	7.1	47.8	

Note: Significance at the 0.01(***), 0.05(**), and 0.1(*) probability levels; [a]The chi2 test if there is a relationship between the gender of the manager of the vegetable plot and the level of commercialization. This ignores the "joint management" category thus the percentage does not add up to 100%.
Source: 2011 field survey data

Generally, control over income between men and women depend on the level of production and income earned. Women are less likely to control high-value vegetables, especially when vegetables contribute a significant proportion to the total household income. High levels of vegetable commercialization of vegetables contribute to male control over vegetable production and revenues. Qualitative information gathered in the field revealed that with vegetables and other horticultural enterprises (such as bananas in Meru) enduring to be key source of household income due to more efficient marketing systems (increased number of traders and organized marketing arrangements), the control of horticultural produce income changed from women to men. Vegetables income has increased quickly surpassing the traditional cash crops, often under the domain of men, such as coffee, tea and pyrethrum.

4.4.3 Constraints hindering the participation of men and women in the commercialization of vegetable enterprises

To understand the challenges facing male and female participation in the commercialization of vegetables and to see how they differ from each other, qualitative work was extended to single-gender focus group discussions. The main challenges faced by women are related to land, labour, and financial capital resources and marketing. Contrary to women, the challenges mentioned by men were more inclined to marketing. Male groups also noted the lack of financial capital as a major constraint.

4.4.3.1 Land resource constraint

In Kenya, land remains the major economic asset that the largest proportion of the population especially rural residents derives their livelihood. Although it is clear that women farmers control the largest proportion of smallholder agriculture, the majority of them do not have legal rights to the same land. Land ownership in Kenya has been controlled by customary laws and traditions, which discriminate against women's ownership of land. Women receive usufruct rights that are limited up to certain transactions. However, a recent amendment of the Kenya National Land Policy (GoK, 2010b) which is aligned with the new Kenyan constitution, advocates equitable access to land, security of land tenure and the elimination of gender discrimination in law, customs and practices related to land and property. The policy also requires appropriate legal measures to ensure that men and women are entitled to equal rights to land and land-based resources in marriage. However, the new law is yet to be internalized and integrated among the customs and traditional laws of the most affected, which are typically in rural communities. The latter societal rules remain strong hence the former is not yet enough to broadly affect the lives rural women.

Therefore, as in most regions of Kenya, the women respondents in the focus group discussions observed that they had no legal rights to land. This weakens their position in terms of making production and marketing decisions. Some women noted that men demand to manage all the proceeds from the sale of vegetables accrued from family land even if they were not involved in the production. One middle-aged woman from Sagana village (Nyeri district) claimed that, "…*my husband rented out all the land including the space where I used to grow vegetables when I refused to share my proceeds with him.*" Usufruct land use rights given to women are withdrawn if women contest their husband's demands. In some cases, women are denied access to water for irrigating their plots. As noted by Dolan (2001) among vegetable growers in Meru district, and recently by Miriti et al.

(2011) among banana growers in the same region, horticultural commercialization has in some regions brought about matrimonial conflict. While men claim returns from their land, women claim for returns from their labour. Such disagreements may result in negative social and economic implications as observed by Dolan (2001). The author illustrates how disagreement between men and women about sharing benefits in Meru district resulted in cancellation of vegetable contracts due to low quality produce. This took place when women deliberately withdrew from the production of French beans because men failed to remunerate their labour contributions. Miriti et al. (2011) observes that in some cases where women have successfully been able to retain income from banana enterprises, their husbands have withdrawn their contribution to the household budget. The empowerment of women through the commercialization of banana production has resulted in some cases in the breakdown of marriages as men resent the fact that women are withholding their banana proceeds.

4.4.3.2 Labour resource constraint

Women work more hours each day than men do, most of which they spend on household chores such as preparing and cooking, fetching water and looking after the children and the old. Across the survey districts, households grow other cash crops such as tea, coffee and rice, which demand significant labour inputs during peak periods such as harvesting. Such tasks are however mainly performed by women. The same applies to vegetable harvesting, a task that is culturally constructed to be a female job. The alternative to family labour is hired labour. Households reported that the cost of casual labour in rural areas has doubled over the last five years due to wages increases. Regrettably, this is not commensurate with the value of returns from farming activities, which have been reduced because of declining farm productivity, and the increasing costs of other farm inputs such as fertilizers, seeds and pesticides. The lack of economic resources thus constrains households from hiring extra labor to substitute some of their work needs. Women indicated that health problems related to irrigating their plots at night and spraying chemicals constrained them from participating in horticultural commercialization. Excess use of chemicals among export market households has significant health implications as discussed by Asfaw et al. (2010a). During the dry seasons when water is rationed, crops must be irrigated at dawn. Cold temperatures occurring during the morning hours are unfavourable to women. This is particularly severe in Nyeri and Makueni districts, where water shortages are most prevalent. It was observed that frequent water shortages in this area had contributed to exit of households from commercial horticultural farming especially among female-headed households.

In addition to farm and household chores, women engage in various off-farm activities such as casual labour in medium and large-scale farms, small businesses such as *dukas,* the roadsides sale of vegetables and fruits, knitting among others. Labour constraints thus limit women from attending agricultural training or participating in community development activities.

4.4.3.3 Financial capital constraint

As demonstrated in the previous section, financial credit is very low given the current abundance of micro-finance institutions in rural areas of Kenya. Traders and PMOs are the most important sources of financial credit for the smallholder vegetable producers in the survey areas. The widespread lack of the collateral - legal land titles- was observed as an important constraint of women's ability to access credit through formal channels. Generally, men and women cited the fear of being unable to pay loans as the reason for not negotiating loans through banks or microfinance organizations. The issue of being unable to pay was mainly attributed to uncertainty inherent to production of vegetables, as they are biologically sensitive to adverse environmental conditions such as drought, pests and diseases, as well as commercial vulnerability to erratic market prices. In addition, most farmers perceived that credit from commercial banks was too expensive. It was also clear during the focus group discussions that women lacked information on existing loans mechanisms.

The lack of capital to purchase farm inputs results in reduced productivity of women's labour in horticultural farming, as they lack the capacity to access and adopt methods of improving soil fertility. Women also typically lack the financial resources to hire manual labour activities such as irrigation and spraying which excludes them from commercial horticultural business. They are also constrained with respect to obtaining modern irrigation equipment such as water pumps and generators. Furthermore, women typically have limited income-generating activity options that could adequately buffer shortfalls in horticultural commercialization performance during "bad" seasons.

4.5 Empirical approaches and results

The descriptive statistics revealed that expenditure patterns of men differ from those of women as well as other gender roles related to intra-household resource allocation and management of vegetable revenues. As elaborated in the introduction section of this study, previous studies suggest that income earned and controlled by women has different effects on household expenditure patterns compared to income earned by men. In this section, the implications of female participation in the commercialization of vegetables on selected household welfare outcomes and the determinants of female participation in the commercialization of vegetables are explored.

Assume a regression model that relates household welfare outcome, $\boldsymbol{Y}$, for household i, and female participation in vegetables commercialization is given as follows;

$$Y_i = \boldsymbol{X}_i'\beta + \alpha_{1i}WP_{ji} + \alpha_{2i}WI_{ji} + \varepsilon_i \quad \mathrm{i} = 1,\ldots,\mathrm{N}; \mathrm{j} = \textit{export market, domestic market} \tag{4.1}$$

where $\boldsymbol{X}$ is a vector of exogenous household and village variables that influence household welfare, outcome $\boldsymbol{Y}$, WP_{ji} is a dummy variable given as 1 if a female member of household i, is the manager of a market vegetable plot j. WP_{ji}, is the share of vegetable income derived from market j that is managed by a female household member and ε is the error term that contains unobserved factors that influence the outcome $\boldsymbol{Y}$. In order to get a better understanding of the implications of female participation in commercialization, a range of household welfare outcomes are considered. These include; per adult equivalent income, per adult equivalent non-land assets, per adult equivalent food and clothing expenditures and household expenditures on school fees and health.

In this analysis, we are confronted with endogeneity problem due to simultaneity bias. This implies that there could be factors that influence $\boldsymbol{Y}$ through $\boldsymbol{WP}$ and $\boldsymbol{WI}$, rendering them endogenous in equation (4.1). Estimating the above model using OLS therefore may result to biased estimates of α_1 and α_2. The endogenous variables can be expressed as follows:

$$WP_{ji} = \theta_i \boldsymbol{Z}_i + \mu_{1i} \tag{4.2}$$

$$WI_{ji} = \theta_i \boldsymbol{Z}_i + \mu_{2i} \tag{4.3}$$

where Z_i is a vector of variables that affect women's production decisions and management of revenues generated from the sale of vegetables. Z_i contains $\boldsymbol{X}$ and other variables that are correlated with female participation in the commercialization of vegetables, but not with household welfare outcomes. In this case, the assumption is that the covariance of error terms for the participation equations (4.2 and 4.3) are correlated with error term of the household welfare outcome. A possible solution is to use an instrumental variable approach. This involves using two-stage least squares (2SLS) where equations (4.2) and (4.3) are estimated in the first stage using a linear or non-linear model and then equation (4.1) is estimated in the second stage. The two-stage least square model relies on the central limit theorem, and is considered robust even with dummy endogenous variables conditional on the identification of valid instruments (Wooldridge, 2002). In this regard, because ***WP*** is a dummy variable, it can be estimated using an IV approach. However, ***WI*** is a functional form model, which is censored at zero since only some female household members manage a certain proportion of income generated from sale of vegetables. In this case, an IV approach may not be suitable. An alternative approach is the Control Function approach. Similarly, to the IV approach, this method involves two stages where the exclusion restriction applies in the first stage. The first stage is estimated using a Probit model for dummy dependent variables to obtain Probit residues (μ_{1j}) and using the Tobit model for censored dependent variables to obtain Tobit residues (μ_{2j}). Unlike the IV approach, in the second stage, the structural model (equation (4.1) is estimated with the observed endogenous variables and the residuals from the first stage as explanatory variables. Equation (4.1) is therefore estimated as:[30]

$$Y_{ik} = \mathbf{X\beta} + \alpha_{1i} WP_{ji} + \alpha_{2i} WI_{ji} + \psi_1 \hat{\mu}_{1j} + \psi_2 \hat{\mu}_{2j} + \varepsilon_i \qquad (4.4)$$

The control variable approach in this context is superior to the IV approach as it also accounts for omitted variables, which may influence female participation but not household welfare outcome. The test for endogeneity is the statistical significance test of the coefficient of the residuals estimated using bootstrapped standard errors (Heckman & Robb, 1985; Wooldridge, 2002;). Control variable approach will be utilized for the analysis in this study.

Identification of the female participation equations

After establishing the appropriate regression method, the next step is to establish the variables that correlate with female participation in the commercialization of vegetables but not with household welfare outcome in order identify equations (4.2) and (4.3). Such variables can also be viewed as instrumental variables (IVs) for female participation. Identification of instruments for female participation in agriculture is difficult and particularly so in the context of rural Africa where most production and marketing decisions are dominated by household heads who are usually male household members. Considering the two indicators used in this

---

30 Addressing endogeneity in health econometric modeling, Terza et al. (2008) find two-stage residual inclusion, which follows same procedure as control function approach consistent, while the two-stage predictor substitution, whose procedure is similar to that of the 2SLS, inconsistent.

study as measures for female participation in horticultural commercialization, they can be related to different instrumental variables used by other authors to measure female bargaining power in rural contexts. It is important to note that these instruments are selected based on the context of the study which may vary from one location to another depending on cultural, social and economic activities as well as geographical and agro-ecological conditions (Quisumbing & de La Brière, 2000). The instruments are also based on the measure of bargaining power being considered.[31] Rainfall patterns for example have been used as instruments for individual household member agricultural income on household expenditure patterns (for example Doss , 2001 in Ghana and Duflo & Udry 2004 in Cote d'Ivoire). Osmani (2007) uses the size of labour, number of dependents, and principal occupation of the household head to instrument female participation in a micro-credit program focused on land ownership and non-land assets in Bangladesh. To establish the relationship between women's assets and shares of household expenditures on children's clothing and education in Bangladesh, Quisumbing and de La Brière, (2000) instrument their current assets along with assets owned at the time of marriage. Further, the authors use various household characteristics such as education of the husband and wife, age, number of siblings, and land holdings of the wife's parents as instruments for women assets at the time of marriage.

Based on these past studies and on the context of this study, various potential instrumental variables are identified. These instruments are related to the production and marketing decisions and various indicators of intra-household resource allocation. For example, women for example are labour constrained due to multiple responsibilities at home and on the farm. As such, we use the size of labour measured as the number of adult females in the household and number of children less than five years old. A large female labour size is expected to positively influence female participation in vegetable plot management and revenue but not influence household welfare outcomes.

Second, we relate market decisions to infrastructure, which is reflected by type of road, distance to the nearest market town and the square of it, and division-level prices for two important domestic market vegetables and two important export market vegetables. These variables are directly related to market opportunities for farm enterprises, which may directly affect women position in the production

---

31 Measures or indicators of bargaining power include income, employment, assets, human capital, decision-making, and other measures such as participation in projects (Doss 2012). In our analysis, the two measures of women's participation in the commercialization of vegetables represent their bargaining power in the commercialization of vegetables.

and marketing decisions relative to that of men. This is based on the observations made from past studies that new market opportunities reduce the role of women in the commercialization of farm enterprises. Maize prices and the size of cultivated land allocated to maize are also included to account for food sufficiency. The inclusion of maize variables, a staple crop in Kenya, is based on the argument that women's participation in market activities is likely to have negative effects on food security and nutrition at household level.

In addition, indicators of human capital that, which could influence women's bargaining power, are included in the analysis. These include the education of the head's spouse and female membership in social groups.[32] The age of the spouse may also influence bargaining power and thus included in the estimation. The primary activity of the household head may have an influence on female participation in horticultural activities. Since the current primary activity may be correlated with the household welfare outcome, we use the primary activity five years prior, which we easily obtain from the dataset collected from the same households in 2005/2006. For export market households, a lagged group membership of the household in a producer and marketing group is also included.

4.5.1 Determinants of female participation in the commercialization of horticulture

Before discussing household well-being implications of female participation in commercialization of vegetables, we report the results from the first stage of the control function approach (equations (4.2) and (4.3) for each market pathway. In the first stage, we use all of the variables, including those hypothesized to influence household well-being outcomes and the instruments used to identify participation equations. The results are given in Table 4.9. The table presents the Probit model estimates of the determinants of female participation in production decisions: management of export vegetable plots (model 1), and the management of domestic vegetable plots (model 2). In addition, Table 4.9 presents the Tobit regression estimates of the determinants of the share of revenue generated from export and domestic market vegetable sales that is managed by female members of the household; export market vegetables (model 3) and domestic market vegetables (model 4).

As mentioned earlier, it is difficult to identify suitable instruments for participation in horticultural commercialization by rural women. Further, the analysis is

32 These variables are included here because they can enhance information access, and thus may influence female participation in vegetable commercialization

Table 4.9: Determinants of female participation in the commercialization of horticulture

	Probit regression estimates				Tobit regression estimates			
	Female export plot manager (dummy) (1)		*Female domestic plot manager (dummy)* (2)		*Female export revenue (%)* (3)		*Female domestic revenue (%)* (4)	
Remittances (1=Yes)	–0.674*	(0.448)	0.564**	(0.248)	22.5	(19.8)	34.53**	(17.1)
Head age (years)	–0.256*	(0.165)	–0.089	(0.088)	–3.39	(7.55)	2.41	(6.9)
Head age squared	0.002*	(0.001)	0.001	(0.001)	0.024	(0.07)	–0.03	(0.1)
Household size(adult equivalent)	0.054	(0.244)	–0.226	(0.158)	–20.39*	(12.7)	–28.2**	(11.6)
Dependency ratio	0.643	(0.519)	0.145	(0.349)	–5.89	(26.9)	10.39	(29.1)
Children attending school	–0.357	(0.390)	0.106	(0.240)	27.67*	(18.8)	1.00	(17.5)
Head education (years of schooling)	0.022	(0.072)	0.057	(0.041)	2.89	(3.36)	2.09	(3.0)
Household major activity in 2010 (1=Non- farm)	1.80***	(0.516)	1.68***	(0.261)	8.21	(20.6)	2.64	(17.3)
Total land owned (acres)	–1.44**	(0.727)	0.027	(0.139)	–38.08*	(25.9)	3.67	(9.5)
Land cultivated (acres)	1.512**	(0.751)	–0.040	(0.170)	41.64*	(27.2)	–0.91	(11.0)
Vegetable area (%total cultivated land)	0.018**	(0.009)	0.008*	(0.005)	0.37	(0.39)	0.57*	(0.3)
Cash crop (1=yes)	0.169	(0.461)	0.670**	(0.286)	–11.43	(22.0)	–10.56	(20.2)
Fertile land (1=yes)	0.105	(0.432)	0.000	(0.260)	40.83**	(19.8)	16.39	(17.7)
Annual rainfall (mm, 2009)	–0.006	(0.005)	0.000	(0.004)	0.13	(0.23)	–0.12	(0.2)
Covariance of variation (monthly precipitation, %)	–0.029	(0.049)	0.037	(0.039)	–0.50	(2.83)	1.04	(2.2)
Shock (dummy)	0.258	(0.435)	0.247	(0.243)	20.21	(18.5)	4.60	(16.0)
Distant to market town	0.405**	(0.191)	0.038	(0.102)	6.68	(7.42)	–0.02	(6.7)
Distant to market town (sq.)	–1.66**	(0.955)	–0.171	(0.503)	–33.30	(37.5)	12.30	(32.8)
Road type (1=good)	0.108	(0.483)	–0.101	(0.274)	7.83	(21.4)	–31.20*	(20.4)

To be continued...

	Probit regression estimates				Tobit regression estimates			
	Female export plot manager (dummy) (1)		*Female domestic plot manager (dummy) (2)*		*Female export revenue (%) (3)*		*Female domestic revenue (%) (4)*	
Female adults	0.681	(0.680)	0.893**	(0.410)	61.55**	(32.3)	59.35**	(29.8)
Spouse age	0.127	(0.222)	0.104	(0.105)	9.65	(9.28)	4.77	(8.1)
Spouse age (sq.)	–0.001	(0.002)	–0.001	(0.001)	–0.08	(0.10)	–0.08	(0.1)
Spouse education	–0.051	(0.080)	–0.078*	(0.046)	–0.21	(3.66)	–3.01	(3.4)
Children below 5 years	0.027	(0.542)	0.186	(0.249)	–24.19	(20.6)	16.23	(16.2)
French beans price(log)	–0.206	(0.570)	–0.107	(0.325)	–12.94	(22.6)	32.34	(23.0)
Snow peas price (log)	–0.632	(0.978)	–0.497	(0.553)	51.11	(40.0)	–35.79	(33.1)
Potatoes price(log)	–2.38**	(1.032)	–0.128	(0.543)	12.09	(49.3)	15.65	(38.0)
Cabbage price(log)	–0.205	(0.680)	0.092	(0.351)	37.27	(28.3)	–51.29**	(24.2)
Maize area (% cultivated area)	–0.023*	(0.013)	–0.009	(0.008)	0.49	(0.55)	–0.36	(0.5)
Maize price (US$/Kg)	–0.161	(0.626)	–0.096	(0.313)	–75.5***	(27.7)	–90.9***	(24.0)
Female social groups(1=yes)	–0.162	(0.238)	0.093	(0.135)	–13.61	(11.7)	5.49	(9.0)
Head major activity(2005) (1=non-farm)	–1.697*	(1.046)	–0.256	(0.385)	–11.27	(34.6)	15.85	(28.2)
Producer group member 2005(1=yes)	0.964*	(0.528)			–21.51	(19.3)		
Female extension contact (1=yes)	1.74***	(0.639)			57.8**	(27.4)		
Female vegetable training (1=yes)	3.07***	(0.809)			6.99	(28.8)		
District fixed effects	Yes		Yes		Yes		Yes	
Constant	4.481	(7.11)	–4.389	(4.36)	–382.4	(329)	–291.6	(296)
Number of observations	242		242		242		242	
LR chi2(38) (export); LR chi2(35) (domestic)	145 ***		92.1***		50.4*		68.0***	
Pseudo R2	0.65		0.32		0.07		0.08	
Log likelihood	–38.9		–97.8		–331.3		–391.686	

Note: Significance at the 0.01(***), 0.05(**), and 0.1(*) probability levels; standard errors are presented in parenthesis. The prices of vegetables are taken at the divisional level
Source: 2011 field data

based on cross-sectional data and a relatively small sample size. This limits the statistical power of the conclusions, which are observed as correlations. Results in Table 4.9 show that female management of export vegetable plots is positively related to the size of land cultivated, the share of cultivated land allocated to vegetables, rainfall and distance to market town. Furthermore, the results support the hypothesis that female access to extension services and training on vegetable production are positive and significantly related to female management of export vegetable plots. World Bank et al. (2009) observe that in countries where agriculture is a major source of employment for women, like in Sub-Saharan countries, capacity building, vocational training and extension services addressing relevant skills and knowledge gaps have been found to be pertinent. The lagged household group membership variable also shows a positive relationship to female management of export vegetable plots.

Female participation in the management of domestic vegetable plots is positively related to the availability of remittances. This can be interpreted as a source of capital for the commercialization of vegetables. As expected, a female household member is likely to manage vegetable plots where the household head's major activity is non-farm. In agreement with the descriptive statistics, female household members are likely to manage vegetable plots if the household is engaged in the production of other cash crops like coffee and tea, which in most cases are under the domain of the male household members. The amount of labour given in the analysis as the number of female adults in the household is also positively related to female participation in the management of domestic vegetable plots. Similar to female management of export vegetable plots, the share of cultivated land allocated to vegetables is positively related to female management of domestic vegetable plots. This is expected since vegetables are labour intensive and most of the activities are carried out by women as demonstrated earlier in Figure 4.2.

The share of revenues from the sale of export market vegetables managed by female members of the household is positively related to the number of children attending school, total land cultivated, total number of female adults in the household, and female access to extension services. The positive association between the size of land cultivated and the share of export market revenues controlled by a female household member could be an indication of women's compensation for their labour input for the production of vegetables and other crops. The same argument could explain the positive estimates of the share of land allocated to vegetable production, although the estimates are not significant. Female access to extension services is expected to provide production and marketing skills that could influence the management of revenues from vegetable sales. On the other hand, the share of revenues from the sale of export market vegetables is a decreasing function of household size, total land owned, and maize prices. This finding agrees with Fischer and Qaim (2012a) who find farm size to be negatively related to the share of banana income that is managed by female household members. Increases in maize prices could be negatively related to the share of revenue managed by females in two ways. On one hand, increases in maize prices may increase the opportunity costs of vegetable production inducing a household to shift labour and land resources to the production of maize and thus reducing income from the commercialization of vegetables. From a different perspective, a bigger share of the jointly managed income could be directed to the purchase of food thus reducing the share of income managed by different members of the household. This later argument, however, holds if the unitary intra-household model is assumed.

The share of revenues from domestic market vegetable sales is positively related to household access to unearned income (remittances), the share of cultivated land allocated to vegetable production and the number of female adults in the household. The positive relationship between land allocated to commercial crop production and female control over revenues agrees with the findings of Fischer and Qaim (2012a) that female controlled income to be positively related with the area under banana production. Large land area cultivated could be related to high female labour input, which may translate into a large share of vegetable income controlled by women. As expected, the price of cabbage, one of the important domestic vegetables is negatively related to the share of revenues from sale of domestic vegetables managed by women. This agrees with the expectation that increases in the profitability of farm enterprises may reduce the role of women even if they were the main contributors of farm production before the change. Female participation in social groups such as village women's groups decreases their share of domestic vegetable sales. Participation in social groups provides women with potential source of income for example ROSCAs (Rotating Savings and Credit Associations). Consequently, such groups may provide an incentive for women to demand less of the income generated from crop sales.

4.5.2 Female participation in commercial horticulture and household well-being

Table 4.10 provides the second stage regression estimates of the control variable approach (equation (4.4). The estimates provided relate female participation in the commercialization of vegetables and household well-being outcomes. The results are provided for four categories of household welfare outcomes measured at per adult equivalent level: non-land assets (mode 1), income (model 2), and food and

Table 4.10: Second stage estimates for the control function model

	Per adult equivalent non-land assets (log) (1)		Per adult equivalent income (log) (2)		Per adult equivalent food and clothing expenditure (log) (3)		Expenditure of fees and health care (log) (4)	
Female control over export vegetable revenue (%)	*0.0000*	*(0.0004)*	*–0.001*	*(0.002)*	*0.0001*	*(0.001)*	*0.001*	*(0.003)*
Female control over domestic vegetable revenue (%)	*–0.0002*	*(0.0004)*	*–0.0005*	*(0.001)*	*–0.001*	*(0.001)*	*0.004*	*(0.003)*
Export vegetable plot manager (1=Female)	*–0.033*	*(0.059)*	*–0.089*	*(0.168)*	*–0.092**	*(0.138)*	*–0.403*	*(0.365)*
Domestic vegetable plot manager (1=Female)	*–0.032*	*(0.135)*	*–0.271*	*(0.352)*	*–0.614*	*(0.452)*	*–1.615**	*(0.916)*
Remittances (1=yes)	–0.010	(0.034)	0.018	(0.100)	–0.077	(0.099)	–0.101	(0.233)
Age of household head (years)	–0.017*	(0.012)	0.003	(0.031)	–0.062**	(0.029)	–0.003	(0.056)
Age of household head squared	0.0002*	(0.0001)	–0.0001	(0.000)	0.001**	(0.000)	0.000	(0.001)
Household size (adult equivalent)	–0.17***	(0.012)	–0.094***	(0.031)	–0.057*	(0.036)	0.272***	(0.077)
Dependency ratio	0.043	(0.045)	0.189*	(0.103)	0.083	(0.114)	0.229	(0.268)
Number of children going to school (below 15 years)	–0.045*	(0.028)	–0.173***	(0.066)	–0.050	(0.074)	–0.158	(0.178)
Head education (years of schooling)	0.006	(0.005)	0.025**	(0.013)	0.038***	(0.013)	0.062*	(0.033)
Head major activity (1=non-farm)	0.015	(0.065)	0.381**	(0.181)	0.358*	(0.226)	0.841*	(0.466)
Total land owned(acres)	0.077**	(0.032)	0.097	(0.070)	0.154*	(0.082)	0.483***	(0.134)
Land cultivated (acres)	–0.047	(0.035)	0.140*	(0.079)	–0.113	(0.083)	–0.43***	(0.153)

To be continued…

	Per adult equivalent non-land assets (log) (1)		Per adult equivalent income (log) (2)		Per adult equivalent food and clothing expenditure (log) (3)		Expenditure of fees and health care (log) (4)	
Vegetable area (% total cultivated land)	0.001	(0.001)	0.004**	(0.002)	–0.0003	(0.002)	0.008**	(0.004)
Cash crop (1=yes)	–0.008	(0.043)	0.144	(0.119)	0.153	(0.127)	0.205	(0.222)
Fertile land (1=yes)	0.042*	(0.025)	0.196*	(0.111)	–0.224**	(0.093)	–0.50***	(0.202)
Annual rainfall (mm, 2009)	0.000	(0.001)	0.000	(0.002)	0.000	(0.003)	–0.002	(0.002)
Covariance of variation (monthly precipitation, %)	0.002	(0.006)	0.006	(0.017)	0.013	(0.024)	–0.034	(0.029)
Shock (dummy)	–0.016	(0.027)	0.186*	(0.099)	0.054	(0.086)	0.508***	(0.186)
ψ (female export plot management)	–0.017	(0.044)	0.104	(0.114)	–0.039	(0.092)	0.099	(0.270)
ψ (female domestic plot management)	0.016	(0.073)	0.100	(0.213)	0.363	(0.260)	0.909*	(0.550)
ψ (female export revenue management)	–0.0002	(0.0004)	–0.001	(0.001)	–0.001	(0.001)	–0.01***	(0.002)
ψ (female domestic revenue management)	–0.0004	(0.0004)	–0.0004	(0.001)	–0.001	(0.001)	0.002	(0.002)
District fixed effects	Yes		Yes		Yes		Yes	
Constant	1.143*	(0.657)	5.765***	(1.887)	3.686*	(2.163)	5.573*	(3.415)
Number of observations	242		242		242		240	
Wald chi2(27)	2837***		442***		141***		270***	
R-squared	0.85		0.52		0.24		0.35	

Note: Significance at 0.01(***), 0.05(**), and 0.1(*) probability levels; Bootstrapped standard errors are presented in parenthesis.

clothing expenditure (model 3). The fourth category of household well-being is monthly household expenditures on school fees and health care (model 4). Non-land assets are an index calculated using Principal component analysis (PCA). The rest of the household welfare outcomes are calculated in monetary values. The coefficients of some of the residual variables are statistically significant implying that female participation is endogenous and therefore it is important to address them. The residuals utilized from the first stage are from all the four models presented in Table 4.9.

From the regression analysis, all of the estimates indicate that female participation in the commercialization of horticulture, both in plot management and revenue appears not to influence per adult equivalent non-land assets and income. These estimates are negative but not statistically significant. The share of income from export vegetable sales managed by female household member yields a positive coefficient for household expenditures on food and clothing, and for expenditures on school fees and health care, but they are not statistically significant. Similarly, the share of female managed revenue from domestic market vegetables is positive but not significantly related to expenditures on children school fees and health care. The positive estimates agree with past studies, which observe that women invest in food and child needs before they can make any investment in physical assets (Njuki et al., 2011). On the other hand, female management of both export and domestic markets vegetable plots show a negative relationship on both food and clothing expenditures and expenditures on school fees and health care. The coefficients for female management of export plots are negative but not significant while the ones for management of domestic vegetable plots are negative and statistically significant. The above findings give an indication of the limited economic returns generated from women-managed enterprises that perhaps could be contributed to by the lack of access to inputs and market information.

4.6 Conclusions and policy implications

This study discusses gender roles and constraints in smallholder commercialization of horticultural farming through both the export and domestic market supply chains. Further, motivated by the hypothetical collective household model, the study has applied empirical models to determine that factors that influence female participation in the commercialization of horticulture and implications on household well-being.

Generally, control over income between men and women depend on the level of production and income earned. Women are more likely to control income from less profitable vegetable enterprises. These are vegetables produced primarily for the domestic market, while men have more control over export market vegetables supplied. Persistent gender disparities constrain women from accessing productive resources as demonstrated using patterns of gender roles in commercialization of horticulture. Women are deprived of access to land, irrigation water, agricultural credit, production and marketing information, and extension services.

Empirical results confirm the hypothesis of this study that female training on vegetable production and marketing, extension services and membership in farmer groups contribute positively to management of export vegetable plots by women. This underlies the importance of skill building in women for ensuring that they do not lose control of those agricultural commodities that are traditionally known to be under women's domain once they become commercialized. Female participation in production and revenue control did not seem to have any influence on the ownership of non-land assets and per adult equivalent income. However, although not statistically significant, the study results agree with previous studies in that the female share of revenue from domestic vegetable sales had a positive relationship with the household expenditures on school fees and health care.

The implications for commercialization-led development of these findings are that gender considerations should be incorporated in the selection of agricultural commodities. This involves identifying commodity preferences and the role of men and women at different stages of the supply chains of these commodities, as well as identifying constraints and opportunities for men and women as relates to intra-household relations. Both men and women should be integrated into vegetable supply chains and provided with gender training to prevent negative intra-household effects of market development and commercialization programmes. For example, this includes organising mixed gender farmer groups as opposed to single-gender groups that result in intra-household conflicts in the management of resources and income. Membership rules of farmer groups and associations should allow multiple members of households to register instead of just one member per household. Commercialization-related projects need to ensure that women do not lose control of crops and income to reduce negative market–household welfare trade-offs as observed during the field survey. In addition, there is a need to educate both men and women on land use rights so that they can integrate them into the existing customs and traditional laws of their rural communities. Access to reliable water sources is vital for female participation in horticulture as demonstrated in this study. In addition to promoting smallholder irrigation schemes, which is taking place in some areas, other farms-level systems such as boreholes and water harvesting should thus be explored.

The empirical model applied has tried to address the endogeneity problems that are likely to give biased results. However, as with many cross-section data sets and with a limited sample size, the findings are only limited to the study area and broad generalization should be discouraged because the results are case specific. Moreover, heterogeneity between regions in terms of the types of marketing channels, socioeconomic conditions, and culture may differ considerably from one situation to another. Further research should therefore explore gender dynamics using panel data, and focus on different types of crops and supply chains to provide better understanding on gender roles and their implications on household wellbeing.

Chapter Five

5. Commercialization of Smallholder High-Value Horticultural Farming: Institutional Arrangements, Transaction Costs and a Cost-Benefit Analysis of Kenya's Vegetable Producers

5.1 Introduction

In recent years, in Africa as in other parts of the developing world, agricultural production systems and their participants have faced significant challenges in the face of changing economic, environmental, and sociopolitical backgrounds. Changing environmental conditions such as increasing unpredictability of weather patterns (extreme temperatures, and precipitation events) and declining productivity of land pose challenges to the agriculture dependent communities (Kirsten et al., 2009; Kristjanson et al., 2012). The globalization of agricultural economies presents farmers with opportunities as well as challenges. While free trade agreements have increased market access for some commodities, they might have negatively affected smallholders, who are forced to compete with both local producers as well as farmers from other countries (GoK, 2012). The globalization of markets has also brought new production and marketing regulations as well as new risks and uncertainties related to volatile global food prices, and financial crises among other global concerns. Consumer demand for traceability of produce to its origin is increasing in order to facilitate monitoring of food quality and safety (Jensen, 2006). Consumer preferences are also changing as income increases pair with rising demand for easy-to-cook food, thus influencing agricultural supply chains.

Marketing exchange arrangements between smallholders and commercial buyers of high-value produce have also been changing to take up the emerging opportunities as well as accommodate the evolving challenges to maintain their position in international markets. In the Kenyan high-value vegetable market, the change in marketing arrangements is evident. Farmers and traders have adopted various strategies to maintain their positions within the supply chains. This study aims at examining the characteristics and the relative costs and benefits of different marketing arrangements, which are unclear. The analysis will be able to examine the incentives for farmers who participate in the respective arrangements and establish the viability of different marketing channels for smallholders. This analysis is intended to somewhat provide indications of why smallholders are exiting from the high-value export market as reported in the recent studies (Okello, 2005; Muriithi et al., 2010).

The next section describes the emergence of marketing institutional arrangements among Kenyan smallholders followed by a brief literature review on types, costs and benefits of institutional or marketing arrangements of high-value agriculture. This is followed by conceptual framework and theoretical underpinning of the study and subsequently the analytical method (cost-benefit analysis). Data and data collection methods are then presented before providing the results and finally conclusion and policy implications.

5.2 Evolution of vegetable marketing arrangements in Kenya

The evolution of vegetable marketing arrangements can be traced together with the growth of Kenyan horticultural industry back to the pre-war colonial period. Minot and Ngigi (2004) describe the growth process of the horticultural sub-sector as complex and rather arbitrary. The author distinguish five periods of growth: the pre-war colonial period, the post-war colonial period, the early independence period, the take-off period from 1974 to 1990 and the 1990 to 2000 period. Up to the present time, two more periods can be differentiated: from early 2000 until 2008 and from 2008 to the present. The early 2000 period is associated with the rapid expansion of supermarkets, especially in Europe (Jaffee, 2003) while 2008 marked the peak of the global financial crisis.

During the *pre-war colonial period*, the commercialization of horticulture was mainly driven by large-scale private farmers (Jaffee & Little, 1994). Participation of smallholder African farmers in commercial horticulture was more visible, although limited, during the *post-colonial period*, when irrigation schemes were developed for collective farming (Minot & Ngigi, 2004). The *early independence period* marked a positive step for the participation of smallholders in commercial horticulture when the government granted them legal rights to the ownership of land that was acquired from settlers (Jaffee & Morton, 1995; Kimenye, 1995; Minot & Ngigi, 2004). This was followed by the *take-off period*, to which where feasible institutional arrangements between smallholders and exporters can clearly be traced. Farmers formed self-help groups to find reliable market for their produce. A good example illustrated by Jaffee and Morton (1995) involved a group of farmers in a Yatta irrigation project that was contracted by the Kenya Horticultural Export Company. The company provided seed and other inputs on credit and subsequently deducted those costs from their proceeds. Similar to the previous period, contract farming was mainly informal and guided by verbal agreements between exporters and the producer groups. The lack of a legal framework to enforce contractual arrangements, however, led to the collapse of most of these marketing arrangements. A major challenge has been "side-selling"[33] through which some farmers evade loan repayment (Minot & Ngigi, 2004).

The period *between 1990 and 2000* presented the Kenyan horticultural sector with new challenges that required the reorganization of smallholder producers. This emanated from increased concerns over food safety, which led to new regulations on food production and processing by vegetable importing countries (Dolan et al. 1999; Minot & Ngigi, 2004). For example, in 1990 importing countries required food retailers and supermarkets to put into place mechanism for ensuring that the food they sold was safe (Dolan et al., 1999). A popular supermarket code of practice introduced for this purpose was the EurepGAP (later changed to GlobaGAP to reflect the diversity in membership). The standard was originally initiated in 1997 by retailers belonging to the Euro-Retailer Produce Working Group (EUREP). In Kenya, to align the industry to these food safety related concerns, the Fresh Produce and Export Association of Kenya (FPEAK) introduced a code of practice (FPEAK, 1999), that provided employment practices, agro-chemicals applications, land use guidelines, and produce documentation to facilitate the traceability of produce (Minot & Ngigi 2004). In early 2000, the EU legislative framework became more focused on food safety. Sanitary and phytosanitary (SPS) issues such as pesticide harmonization processes, certificates of conformity, traceability, and food directives took center stage (Kyengo, 2006). Private sector standards pushed the degree of compliance to every level of production from "field to fork". These included Marks & Spencer, Tesco Natures Choice and Eurep-Gap (Kyengo, 2006).

To safeguard their market position from the new food requirements, market arrangements had to be restructured to mitigate the expected costs of consummating the involved transactions. Existing farmer groups were reorganized to satisfy the need for compliance with these requirements (Jaffee, 2003). Part of this process was reducing the size of groups to less than 30 members (from as many as 350 farmers), to facilitate technical training on those requirements

---

33 "Side-selling" is a common practice in agricultural supply chains that involves by-passing contractual arrangements. Farmers under contract are provided inputs and technical services under the obligation that they deliver the harvest to the contracting firm, often at pre-determined prices. At harvest time, however when the market prices are higher than the contract prices, farmers instead sell all or part of their produce to spot market buyers or other buyers instead of delivering to the contractor.

and to minimize the costs of monitoring widely dispersed farms (Narrod et al., 2009). In practice, compliance with these standards also led to the reorganization of marketing arrangements or contractual arrangements between the smallholder farmers and exporters. Previously informal and verbal contractual arrangements evolved into formal written contracts, somewhat moving to hierarchical market coordination. The main compliance mechanisms among smallholders - Option I and Option II certification schemes- led to two main contractual arrangements; individual and group contractual arrangements, respectively (Asfaw, 2008; Muriithi et al., 2010b). Compliance under the Option I certification scheme involved certifying an individual producer while under the Option II certification; a group of farmers were required to meet standards collectively. In early 2005, compliance with GlobalGAP standards was made mandatory for exporters selling produce to most retailers in Europe. Those requirements, however, have been very dynamic. Their "improvement" implies the adjustment of existing marketing arrangements or changing their governance structure.

The year 2008 marked another direction of the high-value market following the global financial crisis. Although there are a number of factors attributed to the decline in value of agricultural exports (Figure 1.1), such as reduced annual rainfall, the global financial crisis might also have partly contributed to this decline. The global financial crisis and the recent volatility of global food prices have introduced uncertainty for both producers and traders participating in the high-value markets. This required further modifying marketing arrangements and coordination between producers and traders to adjust these uncertainties. Adapting to new market arrangements involves the introduction of new transaction characteristics or improving the existing ones and thus influences transaction costs. Uncertainty of prices, market access, and product quality may also affect transaction characteristics and thus influence transaction costs because of changes or adjustments to marketing arrangements.

5.3 Literature review on the types, costs and benefits of marketing arrangements for high-value agricultural produce

Previous research has produced mixed findings on the costs and benefits of market coordination between small-scale farmers and traders. While some studies show that smallholders have benefited enormously from vertical coordination in the agricultural food industry (Little, 1994; Minot & Ngigi, 2004), others have found that those who actually benefit most are the traders (Fafchamps & Gabre-Madhin,

2001). Benefits accrued from these marketing arrangements may differ from one form of marketing arrangement to another. The diversity is mainly because of the technical requirements of production and the associated production and transaction costs (Simmons et al., 2005). The marketing arrangements, or what is generally discussed as contractual arrangements, have been classified in different ways. Eaton and Shepherd (2001) identify five models of contract farming: -a centralized model, a nucleus estate model, a multipartite model, an informal model and an intermediary model. The *centralized model*, which could describe most of the marketing arrangements between high-value vegetable producers and exporters in Kenya, involves agreement between a trader and a large number of small farmers. These farmers could be organized in a group or deliver produce to a central place where it is collected by the trader. The *nucleus estate* model may not apply to vegetables. It is a variation of the centralized model in addition to the contracted farmers the contractor has their own production facilities for providing inputs to those farmers. The *multipartite model* involves a joint venture between a private company a statutory organization, and the farmers. The *informal and intermediary models*, which have also been evident among vegetable traders in Kenya, have similar characteristics in that both of them are informal contracts. However the latter involves three parties: the main contractor (exporter), a middleman (or broker) who procures and collects the produce and the producers (Odero et al., 2013).

Mighell and Jones (1963) identified three types of contracts based on their objectives with regard to the transfer of decision rights (from the farmer to the trader) and the transfer of risks (Minot, 1986; Simmons et al., 2005; Bijman, 2008). They are distinguished as market-specification, production management and resource-providing contracts. In *market-specification* contracts, the farmer maintains most of the decision rights over farming activities and bears most of the production risks (Bijman, 2008). *Production-management* contracts and *resource-providing* contracts reflect the Kenyan high-value vegetable supply chains in which the contractor controls marketing as well as production processes. In the former, the contractor inspects production processes including the type and amount of inputs used. However, in the latter, the contractor not only specifies the input usage but also supplies the farmer with key inputs (Bijman, 2008).

Lastly, contractual arrangements can also be broadly classified as formal or informal. Fafchamps (2004) observes that in Sub-Saharan Africa, most contracts in agriculture are informal agreements. They are however respected and most of them are generally efficient as a result of self-enforcement by the farmers, as well as the reputation and trust between farmers and traders (Fafchamps & Gabre-Madhin, 2001). In New Institutional Economics (NIE), the set of rules and structures governing the above-described contractual arrangements are referred to as institutional arrangements (North, 1990). They can be formal (for example legal contract enforcement mechanisms) or informal (for example trust, reputation and other self-enforcement strategies).

While the benefits and costs for the above-described arrangements can be analyzed either from the farmer or from the contractor/trader perspective, in this study, the discussion is limited to the point of view of the farmer. Small-scale farmers in developing countries face challenges to increasing productivity, to accessing information about production and market opportunities and they also lack access to financial services (Little, 1994; Fafchamps, 2004). Farmer motivation to enter into contractual arrangements is therefore driven by the potential of these arrangements to address some of these challenges. In his review of contract farming in Asia and Africa Bijman (2008) notes two main benefits of contracting to smallholders (i) to reduce risks in production and marketing and (ii) to improve access to inputs, credit; and technical assistance. In their study in rural Zimbabwe, Masakure and Henson (2005) observe that market uncertainty, income benefits, indirect benefits such as technical farming skills and acquisition of new knowledge, and other intangible benefits as the main factors motivating smallholders to participate in the production of high-value export vegetables under contract. Other benefits of contractual arrangements include access to new and high-value markets, price stability, and the possibility of fixed household income (Little, 1994; Fafchamps & Gabre-Madhin, 2001; Guo et al., 2005; Simmons et al., 2005).

Fafchamps and Gabre-Madhin (2001), however, warn that traders may use contracts to economize on transaction costs thus making the transaction efficient while disadvantaging farmers with respect to the price, quantity and quality of produce delivered. Moreover, private costs may arise if the arrangements require investment in specific assets. Some of the assets may be lumpy investments whose payback period might be significantly long, especially for small-scale farmers. These assets not only attract maintenance costs, but also make production inflexible in terms of the choice of farming activities. In NIE, this aspect is referred to as asset specificity (Kirsten & Sartorius, 2002). The cost of investment and maintenance of those assets can be considered as transaction costs since they are incurred in order to comply with certain institutional arrangements. A good example discussed in the previous section is compliance with GlobalGAP standards. Compliance with these standards involves investing in specific facilities and in some cases hiring technical services.

Different methods have been used to assess the costs and benefits of contractual arrangements. Econometric models such as Heckman selection model have been used to analyze the impacts of contract farming by comparing contract and non-contract participation (for example Warning & Key, 2002: Miyata et al., 2009). Other studies have used gross margin analysis to compare returns from contract farming as opposed to other market arrangements (for example Minot & Ngigi, 2004; Birthal et al., 2005). Most of these studies find positive income impact of contractual relationships. Two main gaps identified from these studies are (i) that they do not differentiate different types of contract topologies that exist among different farmers or group of farmers and (ii) they do not quantify transactions costs, which should essentially be included in gross margin analysis. Assessing the financial performance of these market arrangements is therefore necessary to establish their profitability and viability, which has implications to rural poverty eradication initiatives that focus on such establishments to improve the livelihoods of the poor.

Cost and benefit analysis (CBA) has been used in many other agricultural and development projects to gauge their efficiency (for example Mburu et al., 2007: Bizoza & de Graaff, 2012). However, studies analyzing the feasibility of marketing or contractual arrangements among high-value agriculture activities such as export vegetables are lacking. This study is intended to fill this gap by identifying and quantifying important costs and benefits of different marketing or contractual arrangements, and to carry out a dynamic financial CBA that can shed light on the viability of these arrangements among smallholder farmers.

5.4 Conceptual framework

The conceptual framework is developed from the new institutional economics following Doward's (2001) framework for analyzing institutions. The framework is used in this study to explain the factors that determine transaction characteristics that shape the marketing arrangements between the producers and commercial buyers, and consequently determine costs and benefits accrued from these arrangements. The framework for analyzing institutions draws from two branches of New Institutional Economics (NIE):- the Institutional Analysis and Development (IAD) framework and the Transaction Costs Economics (TCE) framework. The IAD framework has three main elements: the external (exogenous) environment, the action domain and the outcomes (Figure 5.1). The common structure of the two branches of NIE is the external environment representing exogenous variables that influence actor's (farmers and traders) behaviour through different institutional arrangements[34]. The emphasis of the IAD framework is on the behaviour of the actors, which includes the formation of different arrangements guided by trust,

34 Institutional arrangements and market arrangements are used interchangeably in this study. These terms refer to the sets of rules and structures governing contracts and agreements (Kirsten et al., 2009).

rules, communication and bargaining power. The TCE framework emphasizes actors characteristics; for example small farm sizes that influence the need for collective action; the attributes of activities, which include bargaining over the forms and terms of particular institutional arrangements; and the attributes of institutional arrangements (characteristics of different market marketing arrangements).

Following Williamson (1979), this study recognizes that certain transaction characteristics will affect the choice of market arrangements. The author highlighted three components of transaction characteristics that affect the choice of market arrangement: frequency, uncertainty and asset specificity (Figure 5.1). Hobbs and Young (2000) argue that these specific transaction characteristics are the result of certain products (i.e. vegetables in this study context), characteristics which themselves are shaped by external drivers (physical and biophysical factors, policy and governance and socio-economic factors). Changes in transaction characteristics hence alter transaction costs and in that way influence marketing arrangements.

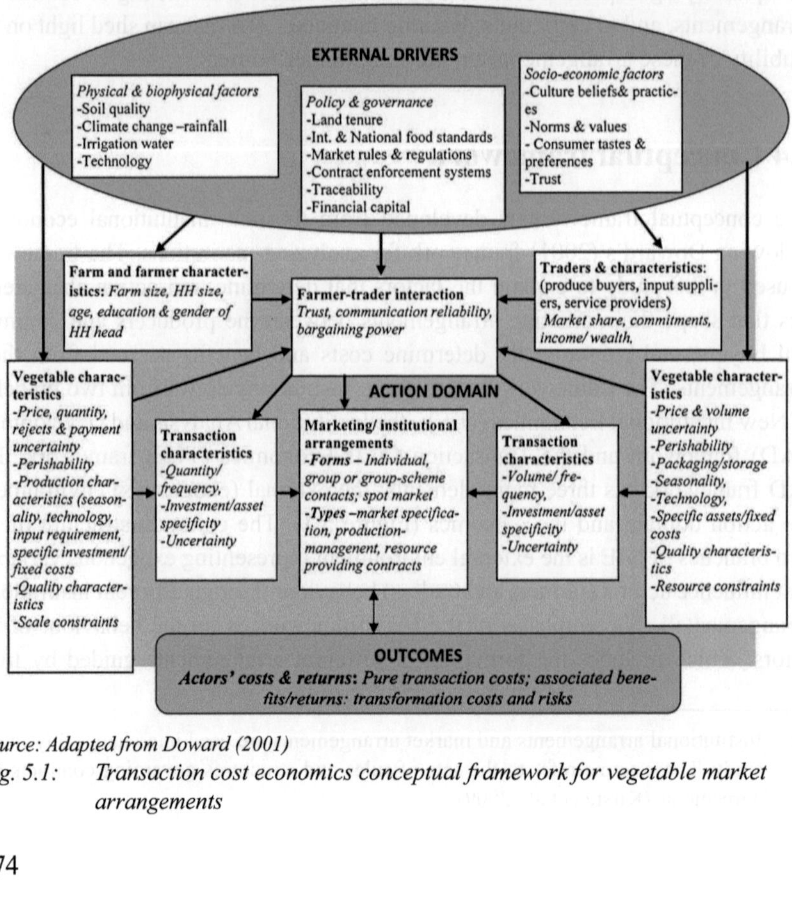

Source: Adapted from Doward (2001)

Fig. 5.1: Transaction cost economics conceptual framework for vegetable market arrangements

Different external drivers can affect vegetable characteristics in various ways and thus influence transaction characteristics. Policy and governance drivers such as private and public, or international and national food regulations as well as traceability may require changes in production (for example type of input used), the quality of produce, and other production technologies that may affect the transaction directly. Climate change (one of the biophysical factors) has implications for seasonal vegetables, which may increase uncertainty over volume supplied, thus influencing marketing arrangements. Changes in consumer tastes and preferences (socio-economic factors) may for example dictate the acceptable quality of vegetables produced. The reorganization of the transactions to incorporate changing external environments is expected to influence transaction characteristics and thus the marketing coordination mechanisms between producers and commercial buyers, and consequently determine the costs and benefits accrued from vegetable enterprises.

5.5 Theoretical foundation

The analysis of this study follows the theory of TCE. TCE recognizes that economic transactions do not occur in a frictionless vacuum; but that buyers and sellers incur costs in conducting a transaction (Hobbs & Young, 2000). Transactions costs can be defined as costs that occur when goods or services are transferred across a technologically separable interface where these interactions occur as a result of technology, division of labour, geographic locations, markets, or people (Sartorius & Kirsten, 2005). These costs are distinguished as *ex ante* and *ex post* costs (Kirsten et al., 2009). *Ex ante* costs include the cost of time, energy, and other resources spent in identifying a suitable transaction partner, identifying product quality, gathering price and other market information among other costs (Hobbs & Young, 2000). These costs are termed information costs. Another set of costs, those that occur during the transaction, are negotiation costs. They include the cost of time spent on negotiating contractual terms, costs of registration (e.g legal registration of farmer groups), and the costs of complying with certain production requirements. *Ex post* are costs incurred after the transaction, are commonly referred to as monitoring and enforcement costs. These are the costs of ensuring that pre-agreed terms of the transaction are adhered to or policing the other party to make sure they honour the terms of the agreement (Kirsten et al., 2009). A good example in the context this study are the costs incurred by commercial produce buyers to monitor contracted farmers during peak demand periods to avoid produce "poaching" by other traders. Institutional arrangements are hypothesized as transaction costs minimizing arrangements that may change with the nature and sources of transaction (Kirsten et al., 2009). Following this understanding, TCE framework is adopted in this study with the recognition that certain transaction characteristics affect the choice of different marketing arrangements through their influence on transaction costs as demonstrated in Figure 5.1. The TCE framework is used to identify transaction costs, which are utilized to determine the viability of different marketing arrangements in addition to production, and marketing costs using CBA.

5.6 Cost-benefit analysis (CBA)

CBA is used as a decision making tool after computing all costs against benefits to come up with a 'net benefit' or a 'net income' (Gittinger, 1982). CBA has been used in many fields to compare the benefits of undertaking an activity with their costs. These include natural resources related projects, farm management, crop production, and other agricultural and non-agricultural related research and development projects (Posthumus & De Graaff, 2005; Tenge & Hella, 2005; Mburu et al., 2007; Bizoza & De Graaff, 2012). There are two types of CBA: economic or social and financial CBAs (Gittinger, 1982). The financial analysis is undertaken from the perspective of a project owner, an investor, or a farmer. It involves the identification and measurement of costs and benefits (usually revenues) measured at market (financial) prices that are directly related to a project or activity. Economic CBA on the other hand measures the economic efficiency of a project, technology or activity from the perspective of society. Net economic benefits are measured as the difference in economic efficiency that is expected to emerge in the economy as a whole with the project compared to that without. Shadow prices are used to calculate the costs and benefits of a project to reflect their opportunity values (Aylward et al., 2000). The objective of CBA in this study, however, is limited to the financial role.

The CBA framework employs a "*with*" *and* "*without*" concept (Gittinger, 1982). This implies comparing one marketing arrangement with certain characteristics against another one that does not have such characteristics. In this study, a comparison is made across different contractual arrangements that differ in certain aspects (individual, group and scheme contracts) and between each of them and spot market exchanges. A financial CBA can be conducted both qualitatively and quantitatively (Macher & Richman, 2008). Qualitative analysis generates the relative profitability of an activity whereas quantitative analysis generates the absolute profitability of a similar undertaking. Past studies assessing the transaction costs of marketing arrangements such as Verhaegen and Huylenbroeck (2001) use the qualitative method citing difficulties in quantifying private costs. In this study, absolute profitability of the different marketing arrangements is determined using a dynamic financial CBA. The advantage of absolute profitability analysis over relative analysis is that it can be used to evaluate the feasibility of different marketing arrangements for smallholders. Further, by quantifying key transaction costs such as the costs of GlobalGap compliance, we are able to shed light on the unclear research concern of whether GlobalGap standards are pushing smallholders out of the export market or whether they are actually catalysts for enhanced market access and thus lead to higher income for smallholders.

A dynamic financial CBA allows the determination of the worth to farmers of making certain investments as required by a certain marketing arrangements. In financial CBA, a net present value (NPV) is obtained by subtracting the discounted costs from the discounted benefits. Discounting translates future costs and benefits into present values. The discount rate applied should represent the opportunity cost of capital expenditures (Gittinger, 1982). A very high discount rate decreases NPV and may result to rejection of a project, which might be a good one, and vice versa. A negative NPV would suggest that engaging in a certain market arrangement is not financially viable because the producers would become worse off in such an arrangement. The formula for calculating NPV is:

$$NPV = \sum_{t=0}^{n} \frac{(b-c)_t}{(1+r)^t} \tag{5.1}$$

where b represents the benefits and c represents investment and recurrent costs. The subscripts and superscripts represent the current and future time respectively, starting from year 0 to year, n, while r represents the discount rate at time, t. An appropriate discount rate for the financial analysis is chosen after considering farmers' opportunity costs of capital. A discounting period is determined based on the current average age of the farmers and life expectancy in the research area.

5.7 Data and methods

The study utilized both secondary and primary data obtained from purposefully selected areas in Kenya where commercial production of vegetables is dominates local economies. These are three districts in the Central province (Nyeri, Murang'a, and Kirinyaga) and two districts in the Eastern province (Makueni and Meru Central). Secondary data was obtained from past studies conducted in this area and from the Kenyan Ministry of Agriculture in the respective districts. Primary data collection was conducted at two levels. The first level involved a household survey that targeted households that had been previously interviewed and trained by the ICIPE in 2005/2006 under a project that assessed the economic impacts of the horticultural sub-sector. This follow-up survey was conducted in June- August 2011 by the author. From the follow-up survey, four vegetable marketing arrangements between farmers and exporters were identified:-individual contracts, group contracts, scheme contracts and spot market exchanges.

Based on these four types of market arrangements, a list of individual farmers (individual contracts and spot market producers) and vegetable PMOs (group and scheme contracts) was developed and ten cases were selected for a third follow-up. Each of the ten cases-studies was organized to incorporate one producer (individual farmer or PMO), a vegetable commercial buyer and an input supplier or service or financial provider directly linked to the producer. Commercial buyers of interest were mainly exporters and middlemen in vegetables export markets, while input suppliers included credit providers and agro-chemical suppliers. The third level of primary data involved in-depth interviews with these actors to gain information on the costs and benefits of different marketing arrangements. The focus however was on the costs and benefits related to the producer. The third follow-up took place in May 2012. Data of interest included production and marketing costs incurred by producers as well as revenues generated from vegetable enterprises. Information on the actors and activities involved in the production and marketing of vegetables and associated benefits and costs was collected. Further, data collection captured the types and characteristics of various transactions between the producers and traders and the associated benefits and costs.

5.8 Results and discussion

5.8.1 Marketing arrangements identified between French bean producers and traders

Although most of the export vegetables have similar marketing arrangements, the CBA conducted in this study does not include all export vegetables produced in the study area. French beans were featured for further analysis for the following reasons. French beans are the most important export vegetables in the study area. Moreover, French beans comprise the largest share of vegetable exports from the country (over 80%) and dominates the vegetables produced by smallholders in the country (HCDA, 2009). In addition, they have limited demand in domestic markets (HCDA, 2009). The contractual arrangements identified in this study involved formal or informal agreements between the transacting parties on certain aspects that relates to the quality, quantity, price, environmental related aspects among others discussed in detail in the consequent paragraphs. Spot market exchange is a form of marketing arrangement that involves establishing a commercial buyer when the crop is ready or nearing harvest. Middlemen (popularly known as brokers) are the main traders that use the spot marketing arrangement. Table 5.1 shows the distribution of different marketing arrangements by survey district. Across all districts, group contract accounts for about 43% of the total sample of households. Specifically, this marketing arrangement is prominent in Kirinyaga (36%), Meru (62%) and Nyeri (54%) districts. Scheme contracts and spot market exchanges are the most common market arrangements in the Murang'a and Makueni districts respectively.

Table 5.1: Type of marketing arrangements by district (% households)

District	Individual contract	Group-contract	Scheme contract	Spot market
Kirinyaga	19	36	23	23
Makueni	0	0	0	100
Meru	12	62	12	15
Murang'a	0	0	92	8
Nyeri	0	54	22	24
Total	***10***	***43***	***26***	***21***

5.8.1.1 Individual farmer contract

This type of marketing arrangement is tailored for individual farmers. It is typically used among medium-scale farmers with the capacity to allot at least one acre of land to French beans production in every season. Further, the farmers must have a permanent source of water for irrigation and the financial capacity to meet certain production and marketing requirements as per the guidelines of the exporter. The agreement is often formal (written) and valid for a period of one year. The contract specifies the type of seeds to be planted, the quality and quantity of product, and expected date of delivery and the price. The trader influences the quality of produce by providing certified seeds. The quality of French beans refers to produce that is not rotten, not infected by pest, and depicts certain colour, texture and shape characteristics. The quantity of produce specified in the contract varies from season to season depending on the market situation and is dependent on the amount of seeds advanced by the exporter. The agreement further specifies the type and brand of pesticides, a schedule of application, and the type and amounts of either organic or inorganic fertilizers to be applied. The cost of seeds is deducted from the farmers' proceeds. In some cases, the farmer requests the exporter to deliver other inputs (i.e. fertilizers and pesticides) or recommends an input supplier. Other guidelines provided include the location of the farm in relation to the surrounding crops or vegetation and the source and drainage of irrigation water. For example, French beans should not be planted on land that had previously been occupied by coffee and/or trees. Most of these requirements are aligned to GlobalGAP or KenyaGAP standards, to which most farmers under this type of contract must comply. In addition, the standards require farmers to provide a facility for handling produce (grading shed), a lockable storage for pesticides and fertilizer, and a cooler for storing harvested produce when necessary. Further, a soak pit for disposing left-over chemicals and contaminated water, clean toilets, harvesting and packaging materials (buckets and crates), and personal protective equipment (chemical handling attire such as gloves, masks, gumboots and overalls), and other equipment such as pumps for spraying agro-chemicals.

Traceability must be ensured. In this regard, farmers are required to keep records of all production and marketing activities including the dates and times of planting, fertilizer and pesticide applications, irrigation schedules and harvesting activities. For this purpose, French beans production farm plots must be labelled. Farmers hire extra assistance for the management of the farm and monitoring production and marketing requirements. The farmer pays the compliance costs of these requirements with little support from the produce buyer. Past studies estimate that exporters contribute about 30% of the initial compliance cost of GlobalGAP standards under individual contracts (Muriithi et al., 2010a). The cost of compliance in some cases is taken into account by companies in negotiation of the contract price. Upon implementing the operational requirements of GlobalGAP standards, the farm is audited and if found compliant, a certificate of compliance is issued. An annual audit however must be carried out during which the certificate is renewed. To monitor the production and marketing of French beans and the correct implementation of production requirements, the exporters have at least one extension agent in each district. The extension agent visits the farmers regularly to supervise production and marketing and offer any technical advice when necessary. Produce is collected from the farmers' premises and payment is made within 1–2 weeks to the farmer's bank account.

5.8.1.2 Group contract

Group contracts arrangement involve a collective agreement made between an exporter and a group of farmers. The group must be legally registered with the Ministry of Culture and Social Services, provide a working committee, and must be willing to engage in the contract without the influence of the trader. For this purpose, a constitution that outlines the rules that guides the group must be provided. The group must also demonstrate the ability to supply collectively the agreed amount of produce within a specified time (weekly or monthly) over the contract period, which is, usually twelve months. Produce quotas are given at the beginning of each season (three seasons each year), which are determined by the quantity of seeds advanced to the group. Group size, in terms of the number of members, may differ from one region to another and in some cases according to the preference of the trader. On average, groups of 30 members are preferred across all districts. In most cases contract are written with most aspects outlined such as the quality and quantity of produce, the date of delivery, the type, brands, and schedule of pesticide and fertilizer applications, as well as the environment surrounding the production plots. In some cases, the traders provide groups with pesticides and fertilizers whose costs are deducted from the group's proceeds. In other situations, the group buys the inputs in bulk and distribute amongst themselves.

In contrast to the individual contract, farmers under group contracts deliver produce to a centralized collection point at their own cost. In the same way as individual contract, farmers under group contracts are required to comply with production and marketing requirements aligned with GlobalGAP standards. However, unlike individual contracts, farmers under group contact make lumpy investments jointly. In addition, group members hire technical staff to manage the group's facilities. These include a group clerk, who assists with grading produce and manages group bank account and a spraying team to assist with the application of chemicals[35]. To ensure that proper production practices are performed at the plot level, farmers must erect spraying stands, observe farm hygiene protocol (including clean toilets and bathrooms), and obtain the required harvest and produce handling facilities (plastic buckets and polythene bags). Farmers are also required to label field plots to facilitate traceability. However, in most cases traceability of produce to the plot level under group contracts is complex because of the small amounts of produce contributed by each group member to the collective group basket. Unlike individual contracts, farmers under group contracts receive a group GlobalGAP certificate that in most cases is retained by the trader. This way the trader secures his supply by reducing the mobility of the group in terms of selling to other commercial buyers[36]. Similar to the individual contracts, the trader

35 The incidences of pesticide-related health impairment have increased among smallholder farmers in Kenya (Asfaw et al., 2010).

36 This portrays a lack of trust between the contract parties. The trader typically distrusts the farmer as the custodian of the certificate. Since the certificate enhances market access for the produce, the trader fears that farmers will use it as a tool to

employs extension agents to monitor farmers for correct production management and side-selling. The agents supervise critical aspects of the production process such as pesticide application and the general maintenance of good agricultural practices. During the harvest, farmers are visited more than once a week to ensure they deliver all produce to the trader. An agreed amount of produce is collected three times a week. Payment is made 1–2 weeks after collection of the produce to the group's bank account. Payments are then distributed to the respective members (cash or bank account) based on the quantity of produce delivered less agreed group expenses.

5.8.1.3 Scheme contract

Scheme contracts are a flexible form of marketing arrangement in which a trader establishes up a trading center and invites farmers in the area to produce French beans and deliver produce to the center. The farmers are selected based on their willingness to commit at least 180 meters square of land (1 unit) to French bean production for at least one planting season. The participating farmers must also have access to irrigation water and an operational bank account. Unlike the individual and group contracts, in scheme contracts farmers are provided with most inputs (seeds, fertilizer and pesticides) whose cost is recovered from a share of the output.[37] The provision of seeds enables the trader to determine the type and quality of produce. A formal contract that specifies the amount of inputs provided, the expected output, and the price is issued to individual farmers by company agents. While this may appear similar to individual contracts, they differ in the scale of production, the stringency with which good agricultural practices need to be

---

divert produce to other commercial buyers awarding better prices for the produce. The lack of trust in contract enforcement been cited as a major cause for contract failure in developing countries (Sartorius & Kirsten, 2007). In their study on finding missing markets on export crop adoption and market intervention in Kenya; Ashraf et al., (2009) observed that the lack of trust that farmers will make specific investments and that exporters will purchase their produce as a contributor to the decrease of smallholder participation in the export market.

37 For one unit plot (180m²), farmers receives 400 grams of seeds, 2 Kg of DAP fertilizer, 2 Kg of CAN fertilizer, a pair of gloves, and five agro-chemical applications. The total cost of this package is approximately Ksh.750 and is repaid with 25 Kgs of French beans (25kg*Ksh.30). Given the high risk of losing inputs as they depend on loose seasonal contracts, traders restrict the number of units that can be planted by each farmer to five.

applied and in provision of inputs.[38] Scheme contracts also tend to be smaller in scale and less stringent on good agricultural practices compared to individual contracts. In addition, scheme contracts are only valid as long as the farmer is producing with supplied input materials. This is approximately one season (three months).

Unlike the group contracts, this arrangement lacks any group leadership aspect given that each farmer deals with the buyer on an individual basis. However, in the same way as the group contacts, farmers bring their produce to a collection center at their own transportation costs. In contrast to the previously discussed contracts, farmers under this arrangement are not required to make lumpy investments in facilities required for GlobalGAP compliance. Nevertheless, they must engage in basic good agricultural practices such as adequate farm hygiene, which includes clean toilet facilities, pesticide disposal pits, and personal protective equipment. Produce is delivered three times a week and, payments are made 2–3 weeks after deliveries to the bank account of the farmer. The deduction for the inputs is made from the first delivery proceeds. To monitor implementation of the production and marketing procedures, the company has about six employees at each center that serve an average of 60 farmers. Three of the employees are permanently on the company payroll, including a center controller who guides the process of awarding contracts, a grader who determines the quality of delivered produce, and a field assistant who provides technical advice to farmers and monitors them during the harvesting period to avoid side-selling. The other three employees form the spraying teams that work on a temporary basis during the growing season.

5.8.1.4 Spot market exchange

The spot market exchange mostly involves selling produce to brokers or small-scale and transient exporters, commonly referred to as "briefcase" exporters. Farmers that engage in this arrangement are typically part-time producers who enter and exit production according to the market situation. In that way, they only participate during periods of peak demand for temperate vegetables,

38 The commercial buyers endeavor to create trust by translating contract documents into the national language (Kiswahili) which most of the farmers can read and write. In addition, center employees are recruited from local communities to facilitate communication with the farmers in their local dialects.

mostly during the European winter season. Unlike the other market arrangements where there are prior production and marketing agreements, in spot market exchanges the farmer contacts the buyer once the produce is ready for harvest. This arrangement may be considered efficient, as conditions approach those of perfect competition and prices are determined by the market forces. In most cases, farmers selling through this arrangement do not comply with most of the GlobalGAP requirements. However, they might be required to demonstrate compliance with some requirements related to good agricultural practices such as high quality produce and farm hygiene.

The first three market arrangements may be regarded as innovative marketing arrangements and are generally thought to be more financially rewarding than the spot market exchanges. However, this may not be the case. Establishing such arrangements may not only require changing production and marketing practices, but also making lumpy investment in facilities to comply with certain market requirements that create higher costs. These costs can be regarded as transaction costs. They consist of information and search costs, negotiating costs to reach contract agreements (on price, quantity, date or season of delivery), and monitoring costs to ensure that agreements are honored. Negotiating costs also include the costs of complying with market requirements that are dictated by the transaction partners. On the other hand, spot market exchanges are highly uncertain as they are based on market speculation. In the absence of perfect market information, this arrangement is less efficient than contractual arrangements.

5.8.2 Consideration of different costs and benefits

Table 5.2 shows a summary of costs and benefits considered in this analysis. The situation of farmers producing based on contractual arrangements is compared with the spot market exchange as a reference. Benefits, direct costs and transaction costs generated by the each contractual arrangement can be higher (+), lower (–) or similar (=) in comparison with those of the spot market exchange. In addition, uncertainty about price, quantity sold, payment, quality and input supply are determined. To be able to make inferences into the different marketing arrangements, further analysis include only those districts where at least three of the marketing models were being practiced (see Table 5.1). The context of the marketing arrangement may be different for each district thus it is important to differentiate them by district. Consequently, the Kirinyaga, Nyeri and Meru districts are considered.

Table 5.2: Potential costs and benefits of different French beans marketing arrangements in comparison with spot market exchange and their hypothesized relative levels

	Individual contract	Group-contract	Scheme contract	Spot market**
Benefits				
Price	+	+	+	
Quantity	+	+	+	
Direct costs				
Production costs	+	–	–	
Market related production costs	=	+	+	
Transaction costs				
Investment cost				
Investment in GlobalGAP standards	+	+	+	
Setting up group	n/a	+	n/a	
Recurrent cost				
Information costs	–	–	–	
Negotiation costs (including GlobalGAP standards)	+	+	+	
Monitoring cost	=	+	=	
Uncertainty				
Price change	–	–	–	
Quantity sold	–	–	–	
Delayed payment	+	+	+	
Nonpayment	–	–	–	
Can negotiate on price	–	–	–	
Rejects/quality	+	+	=	
Input supply and crop failure	–	–	–	

Note: The signs indicate the expected direction of the changes in revenues, costs and uncertainty (higher (+), lower (–) or similar (=) of different contractual arrangements relative to spot market exchange which is the reference of this comparison

The benefits arise mainly from French beans yields. This is captured as gross value product, which is the quantity multiplied by the contract price or farm gate price for farmers selling through the spot market exchanges. For the contractual arrangements, the *price* of French beans is largely assumed better than through spot market exchanges, which may not actually be the case.

Generally, individual contractual arrangements offer the highest prices across the three districts (minimum of Ksh.50/kg of French beans). While the minimum legal pricevia group contract is about Ksh.40/kg of French beans (GoK, 2011), the net contract prices is highly variant as it is established based on the proceeds received in the group's bank account after deducting the value of rejected produce. In some cases, farmers quoted price as low as Ksh.25/kg. On the other hand, scheme contracts receive a constant contract price that ranges between Ksh.30–35/kg. Spot market exchange prices vary by season and may be higher or lower than all other contract prices depending on prevailing market conditions. During periods of peak demand, brokers typically offer higher prices than contract prices. Spot market exchange prices were quoted between Ksh.20–.100/kg of French beans. Inflexibility of the contract prices during peak periods invite farmers under contracts to "side-selling", an issue that has been recognized as a major threat to the high-value vegetable industry (GoK, 2011). Exporters interviewed during the survey indicated that side-selling was the major reason cited for the termination of smallholder contracts. Under a newly established law "HCDA legal order 2011", the government recommends registration of all actors including exporters, brokers, and producers (of all scales) as part of a permanent solution to side-selling (GoK, 2011)[39]. Successful implementation of this law is expected to eliminate the long-time challenge of produce 'poaching' that has persisted since the post-colonial period.

The *quantity* of French beans sold across different contractual arrangements per year is also expected to be higher than for the spot market exchange. While most of the contracted farmers produce up to three plantings year round, spot market producers mainly operate during peak demand periods with an average of two crops per year. Contract farmers are also likely to have higher yields because their produce is more closely monitored by technical staff employed by the traders. Technical services such as spraying, produce scouting and advice on pest and disease control, are provided by traders, whereas spot market exchange farmers lack access to such technical services.

The main cost components are direct and transaction costs. Direct costs include production and marketing costs transaction costs are differentiated as investment and recurrent costs. Important p*roduction costs* include farm inputs such as seeds, pesticides, insecticides and fertilizers, and land preparation and labour costs (both

39 Side-selling per se is not a problem to the government as horticultural sub-sector has had impressive growth, that is often attributed to minimal government intervention. However, the motivation for the government to control side-selling is to protect the image of the country in the international market by ensuring that exported produce is of high quality and that it contains the minimum acceptable levels of pesticides. Controlling side-selling is expected to enhance the adoption good agricultural practices and improve the traceability of produce to the plot level.

family and hired labour) incurred during planting, weeding, irrigation, chemical application and harvesting. Under the group contract arrangement, farmers purchase inputs in large quantities and thereby benefit from lower prices. Similarly, scheme contract producers receive inputs from the exporter at slightly less than the market prices. In contrast, individual contract producers purchase inputs both individually and occasionally from supplied by exporters upon request. Furthermore, individual contract farmers mainly use hired labour while other arrangements use both hired and family labour. As such, production costs for the individual contract arrangements are likely to be higher than all the other market arrangements.

Market production costs include harvesting costs, the costs of transporting produce to delivery centers, sorting and grading costs, and the costs of establishing marketing facilities. Market production costs are expected to be higher in the group contracts and scheme contracts compared to the individual contracts and spot markets, since the former must transport their produce to collection centers at their own cost. In contrast, produce from individual contract and spot market exchange producer is collected by the contract partner.

Investment transaction costs include the *ex-ante* costs of the contracts. The major investment costs in the contractual arrangements are of the measures taken to comply with GlobalGAP standards, including various buildings and facilities. In addition, group contract producers incur extra costs of group registration and for measures taken to comply with the guidelines of the group. *Recurrent transaction costs* include the costs involved in maintenance of activities required for compliance with standards and the contract. These entail the costs of renovating facilities, replacement of product handling gear (gloves, masks, and boots), record keeping, annual auditing costs (including annual soil and water analysis and standard certificate renewal). Farmers in either of these marketing arrangements are thus expected to incur higher transaction costs compared to spot market exchange participants. Information costs include the cost of searching for market information and for potential commercial buyers. Spot market producers are likely to incur higher recurrent market transaction costs, as they have to search for potential buyers at their desired price. In the absence of perfect marketing information, those producers are likely to incur significantly higher recurrent costs than contractual arrangement producers. Monitoring costs for the spot market are expected to be minimal as many transactions are done at the point of purchase.

Uncertainty regarding the price and quantity of supply is lower for all the contractual arrangements compared to the spot market. In contrast to contractual arrangements, spot market exchange does not require specifications of future supply, time, and expected price. Spot market production is based on speculation of market conditions. Another aspect of uncertainty is the delay of payments or non-payment.

Although payment across all the three contractual arrangement is ideally supposed to be made within 2–3 weeks of produce delivery, farmers producing under each arrangement cited delayed payment as a major challenge of trading with exporters. Delayed payments intensify side-selling, especially among group contract producers whose payment procedure is longer relative to the other two contractual arrangements. Spot market producers on the other hand are mostly paid in cash. Immediate need for money compels contract producers to "cheat" and sell part of their produce to brokers. Some farmers reported that bank payment procedures are inefficient because of the time lag for cheque deposit funds to become available. Mobile phone payments are still limited, but gradually gaining popularity especially among spot market producers as an alternative to cash payments. Securing a contract is ideally expected to minimize the risk of non-payment. Legally, one or two government representatives from the Horticultural Crops Development Agency (HCDA) and/or the Ministry of Agriculture should act as a witness during contract signing (GoK, 2011). This is intended to minimize future cases of strategic default and opportunistic behavior from either of the contract parties. The spot market exchange does not facilitate this legal aspect of trading and is therefore more likely to be affected, especially when the produce is sold on credit.

Uncertainty can also be related to produce quality. A significant amount of French beans is rejected due to unacceptable quality characteristics. Over 60% of the farmers across all marketing arrangements reported rejected produce, with the highest percentage of rejects reported by farmers under group contract (over 30%). Produce rejects can be classified into two types: grading-shed (or farm-gate) rejects, which are related to the physical characteristics of the produce and factory rejects attributed to product qualities such as chemical residue levels. Individual and group contract producers incur both kinds of rejects, which are not often compensated for. In contrast, scheme contract and spot market producers do not incur additional rejects beyond the grading step.[40]

5.8.3 Socio-economic characteristics of households by marketing arrangements

Table 5.3 shows characteristics of households based on marketing arrangements. The average age of the household heads (47.8 years) indicates that the sample population is generally of active age. Taking into account that vegetable production is

40 From a different perspective, spot market producers may incur higher rejects because they lack understanding of future supply and demand with respect to product quality.

Table 5.3: Household characteristics by marketing arrangements

Marketing arrangement	Head gender (male=1)	Head age	Head education	Household size	Land owned (acres)	Land cultivated (acres)	Vegetable area (% of cultivated land)	Number of vegetables produced
Individual contract (n=14)	1.0	47.5	10.1	5.3	2.5	2.3	38.19	2.21
Group contract (n=60)	0.9	46.0	9.2	5.5	2.3	2.1	60.11	2.80
Scheme contract (n=36)	1.0	50.8	8.3	6.0	2.1	1.8	44.71	2.19
Spot market (n=29)	0.9	47.1	8.3	4.6	2.1	2.0	56.89	2.31
Total sample (n=139)	**0.9**	**47.6**	**8.8**	**5.4**	**2.2**	**2.0**	**53.24**	**2.48**
t-Test for significance of mean difference between farms with contracts versus spot market								
Individual contract versus spot market		0.362	1.8*	0.664	0.107	0.238	–18.7**	–0.096
Group contract versus spot market		–1.155	0.891	0.859*	0.199	0.037	3.22	0.490
Scheme versus spot market		3.612	–0.041	1.431**	–0.314	–0.233	–12.18	–0.116

Note: Significance at and 0.01(***), 0.05(**), and 0.1(*) probability levels

labour intensive, a household with an average of five adults can harvest one acre of French beans within a half a month or 18 person days (Minot & Ngigi, 2004). However, farmers usually hire harvest labour due to the limited shelf life of French beans. Farmers have small average land sizes (mean 2.2 acres) most of which is cultivated (mean 2.0 acres). Vegetables take a substantial portion of cultivated land ranging between 38% (among individual contract farmers) to 60% (group contract farmers).

Table 5.3 also shows diversification in regards to the number of commercial vegetables produced. All households produced at least two commercial vegetables. Considering that vegetable farming is a highly risky enterprise, spreading the risk by investing in more than one enterprise is common among to farmers. A two-sample t-test with unequal variances was used to determine the differences in the characteristics between farmers participating in contractual arrangements and those selling in the spot market. The test results shows that individual contracts households have significantly higher education (average 10.1 years) than spot market households (8.3 years) ($p<0.1$). The result also indicated that group and scheme contract households were large than spot market households.

5.8.4 Plot level financial CBA of marketing arrangements

An overview of revenues, investment costs, and production costs, and gross margins used for financial cost-benefit analysis for different marketing arrangements; are presented in Table 5.4, Table 5.5 and Table 5.6 for the Kirinyaga, Meru and Nyeri districts respectively. In the study area, French beans production is feasible for a maximum of three seasons under good market and production conditions. Uncertainty of demand and product prices, especially during off-peak periods, precludes some farmers from year round production. Furthermore, climate related challenges such as water shortages during the dry seasons and excessive rainfall (the Meru district is particularly affected by flooding) force farmers to abandon production during certain periods of the year. In this regard, the analysis examines two scenarios; two-season and three-season production. Farmers participating in the spot market exchange typically participate in the market for only one or two seasons, mainly during peak periods of the year. As such, only the two-season scenario is considered for the spot market producers.

As this is a financial analysis, farmer benefits include the gross value of total quantity of French beans sold. The total quantity harvested per year is valued, less the rejected produce. The value of rejects is not accounted for, as they have no market value. This was difficult to compute because not all rejected produce is returned to the farmers. Subtracting the rejects from the sales totals is important for the analysis because different arrangements have different levels of rejects.

Table 5.4: Overview of investment costs, revenues, operating costs and gross margins of French bean production in Kirinyaga district

French bean production (per acre, per year)	**Two-season scenario**				**Three-season scenario**			
	Individual contract	Group contract	Scheme contract	Spot market	Individual contract	Group contract	Scheme contract	Spot market
Average yield(Kg)	4160	4000	3833	3670	6240	6000	5750	3670
Average price(Ksh/Kg)	50	40	30	39	50	40	30	39
Gross Value product (GVP)	208,000	160,000	115,000	143,130	312,000	240,000	172,500	143,130
Production costs	102,380	88,913	84,720	100,043	153,570	133,370	127,080	100,043
Marketing costs	2563	2203	3687	1095	3845	3305	5530	1095
Transaction costs								
GlobalGAP investment	85575	37962	16725	6700	85575	37962	16725	6700
Additional investment costs	1500	4200	1500	1500	1500	4200	1500	1500
Recurrent transaction costs								
GlobalGAP costs	30139	20970	7920	5600	30139	20970	7920	5600
Additional recurrent costs	1122	4877	987	1000	1122.05	4876.6	987.35	1000
Total operating costs	136,204	116,963	97,314	107,738	188,676	162,522	141,517	107,738
Gross margin	71,796	43,037	17,686	35,392	123,324	77,478	30,983	35,392
Labour productivity (GVP/labour cost)	3.5	3.3	2.4	2.7	3.5	3.3	2.4	2.7
Return to investment (GVP/Inv. cost)	2.4	3.8	6.3	17.5	3.6	5.7	9.5	17.5

Note: US$1=Ksh.79

Table 5.5: Overview of investment costs, revenues, operating costs and gross margins of French bean production in Meru district

French bean production (per acre, per year)	Two-season scenario				Three-season scenario			
	Individual contract	Group contract	Scheme contract	Spot market	Individual contract	Group contract	Scheme contract	Spot market
Average yield(Kg)	4200	4040	3867	3700	6300	6060	5800	3700
Average price(Ksh/Kg)	55	50	35	40	55	50	35	40
Gross Value product (GVP)	231,000	202,000	135,333	148,000	346,500	303,000	203,000	148,000
Production costs	107700	95550	91500	110780	161550	143325	137250	110780
Marketing costs	2660	2297	2520	780	3990	3445	3780	780
Transaction costs								
GlobalGAP investment	87287	38721	17060	6714	87287	38721	17060	6714
Additional investment costs	1700	2100	1500	1500	1700	2100	1500	1500
Recurrent transaction ccosts:								
GlobalGAP costs	28350	25079	8350	5400	28350	25079	8350	5400
Additional recurrent costs	1220	4585	700	1000	1220	4585	700	1000
Total operating costs	139,930	127,511	103,070	117,960	195,110	176,434	150,080	117,960
Gross margin	91,070	74,489	32,263	30,040	151,390	126,566	52,920	30,040
Labour productivity (GVP/ labour cost)	3.5	3.6	2.5	2.4	3.5	3.6	2.5	2.4
Return to investment (GVP/Inv. cost)	2.6	4.9	7.3	18.0	3.9	7.4	10.9	18.0

Note: US$1 =Ksh.79

Table 5.6: Overview of investment costs, revenues, operating costs and gross margins of French bean production in Nyeri district

French bean production (per acre, per year)	**Two-season scenarios**			**Three-season scenarios**		
	Group contract	Scheme contract	Spot market	Group contract	Scheme contract	Spot market
Average yield(Kg)	3920	3760	3600	5880	5640	3600
Average price(Ksh/Kg)	40	30	42	40	30	42
Gross Value product (GVP)	156,800	112,800	151,200	235,200	169,200	151,200
Production costs	85433	81290	96550	128150	121935	96550
Marketing costs	2530	2563	705	3795	3845	705
Transaction costs						
GlobalGAP investment cost	38283	17371	6536	38283	17371	6536
Additional investment costs	5450	1500	1500	5450	1500	1500
Recurrent transaction costs						
GlobalGAP recurring costs/year	21075	8140	5770	21075	8140	5770
Additional recurrent transaction costs	4120	1520	1000	4120	1520	1000
Total operating costs	113,158	93,513	104,025	157,140	135,440	104,025
Gross margin	43,642	19,287	47,175	78,060	33,760	47,175
Labour productivity (GVP/labour cost)	4.1	2.5	3.0	4.1	2.5	3.0
Return to investment (GVP/Investment cost)	4.2	6.0	18.8	6.3	9.0	18.8

Note: US$1r=Kshs.79

The information presented in Table 5.4 supports the earlier conclusion that individual contract producers have higher yields and prices compared to all the other marketing arrangements. This applies to both two-season and three-season scenarios. Under scheme contracts, considering two-season scenarios, farmers in Kirinyaga district received less Gross Value Product (GVP) (Ksh. 115,000) than spot market producers (Ksh.143,130). Production costs comprise the highest operating cost component (over 80%). Detailed information on the different aspects of the production costs included in the analysis is given in Appendix 5A (tables 5A-1, 5A-2 and 5A-3). Under the two-season scenario individual contract and spot market producers incurred higher costs of production than to the other two marketing arrangements. Scheme contract producers report lower gross margins than spot market producers in both two-season and three-season scenarios. Marketing costs vary across marketing arrangements. Scheme contracts have the highest cost for both two-season and three-season scenarios.

In Meru district, results are similar to those of Kirinyaga district (Table 5.5). Scheme contract producers receive the lowest price and report less GVP than spot market producers under the two-season scenario. Production costs are the most important operating costs across all marketing arrangements. GlobalGAP compliance recurring costs represent a substantial share of the operating costs among the individual and group contracts (with 20% and 15% of total operating costs for two-season and three-season scenarios respectively). In contrast to Kirinyaga district, all of the contractual arrangements report higher gross margins than the spot market for both scenarios.

In Nyeri district scheme contract producers have the lowest price (see Table 5.6). Considering two-season scenarios, this arrangement reports the lowest GVP. Production costs are the most important operating cost, followed by GlobalGAP recurrent costs. The spot market exchange seems more profitable than both the group and scheme contracts in the two-season scenario. This further applies to scheme contract under the three-season scenario.

Generally, across all the three districts, spot market producers may have incurred higher input costs per season. Unlike contract producers who receive subsidized seeds from traders, spot market producers purchase retail seeds at typically higher prices. Expenditures on fertilizer are the same across marketing arrangements as fertilizer is currently subsidized by the government (Appendix 5A).[41] Differences

41 The government initiative to subsidized fertilizers costs began in 2010. The initiative boosts food security through improving of farm productivity by providing access to cheap inputs. The cost of a 90kg bag of DAP fertilizer has been reduced

in labour inputs reflect the differences in labour costs across the districts. Labour costs are highest in Meru district with an average of Ksh.220 per man-day, while the average costs in Kirinyaga and Nyeri are about Ksh.190 and Kshs.170 per man-day respectively.[42] A possible explanation for this difference is the greater production of other cash crops such as coffee and *Khat* (locally known as *mirra*) in Meru district that creates competition for labour resources especially during harvesting periods. The largest proportion of transaction costs across all the contractual arrangements was for action taken to comply with GlobalGAP standards. The category of "other investment costs" includes legal fees for irrigation water paid by all producers and group registration costs for group contract producers. The costs of maintaining compliance with GlobalGAP standards represent highest proportion of recurrent transaction costs for all contractual arrangements. This includes the costs of repairing buildings, facilities, and equipment, annual soil and water analyses, certification renewal, and record keeping costs. Other recurrent transaction costs are related to the search for market information, contract negotiations and monitoring, land rental (mainly for group contract producers that use hired produce handling and grading facilities-), renewal, and the certification renewal, and the incidental and opportunity costs of time spent participating in meetings and acquiring extension training.

The gross margins for French beans production presented in Tables 5.4–5.6, are lower than those estimated in past studies. For example, Minot & Ngigi (2004) for example estimate gross margins of Ksh.76,733 per acre for this crop in Kenya. The authors however do not take into account the implicit costs associated with family labour as well as transaction costs related to compliance with GlobalGAP standards. A closer examination of gross margins reveals that scheme contract producers have lower gross margins than spot market exchange producers. It is important to note here that although the spot market seems more lucrative, it is associated with higher market uncertainties. Spot market producers do not comply with GlobalGAP standards and thus face higher probability of product rejection or a lack of buyers for their produce. Although GlobalGAP compliance was made compulsory for most of the exporters in early 2005, the lack of government enforcement of these standards may contributed to export market access for

from Ksh.4000 to Ksh.2840, while the cost of a 90Kg bag CAN was reduced from Ksh.2300 to Ksh.1500.

42 The daily rates for unskilled labour have tripled over the last ten years in Kenya, from an average of US$1.2 in 2001 to about US$3 in 2011 (USAID;2012); which may reflect inflation and the education of the workforce.

uncertified products, which are mainly sold through the spot market exchanges.[43] However, introduction of the "HCDA legal order 2011" may eliminate this marketing arrangement since all vegetables producers are required to register with the respective contracting companies and observe good agricultural practices aligned with GlobalGAP standards (or equivalent- KenyaGAP standards) (GoK, 2011). Additionally, the law requires that all the traders (including brokers) register the sources of their produce.

5.8.5 Financial cost-benefit analysis

Financial CBA is used to examine the profitability of the different marketing arrangements. This involves an absolute profitability analysis that can be used to evaluate the feasibility of contractual arrangements for marginal farmers especially considering the high demand for compliance with new market requirements, escalating input and unskilled labor costs. The benefits of contractual marketing arrangements have been perceived as being an effective link to the market by farmers themselves and more so by the government and development agencies. Access to a ready market for produce is the key motivation for the adoption of such contracts. In this analysis, it is assumed that the net present value (NPV) for each of the contractual arrangements will be positive and higher than that of the spot market. A discount rate of 14% per annum and a period of 22 years were established for CBA analysis. The discount rate is calculated as the mean monthly lending rate that applied by local savings and credit schemes to farm loans over the previous five years (2006–2010). The discounting period length of 22 years is determined after considering the average age of the farmers participating in the commercialization of export vegetables (48 years according to the household survey, see Table 3.2 & Table 5.3) and general life expectancy in the study areas (about 70 years). It is thus expected that farmers on average will continue cultivating French beans for the next 22 years. The life span of most facilities is set as 7 years following Asfaw (2008). Based on this estimation, it is assumed that

43 The other possibility is laxity among retailers (importers) in ensuring that all products are certified. During the high demand seasons, retailers purchase produce from any exporter including the seasonal exporters who purchase uncertified produce from brokers and spot market producers. This was reflected in an interview conducted with the a manager from one of the export companies who claimed that , "*…the retailers in Europe who ask for GlobalGap certification are the same ones who receive uncertified produce during the high demand seasons…*" Interview, Senior manager, Frigoken exporter, 28/05/2013.

investment costs are incurred every 8th year and with an annual increase of 10% to account for inflation. Recurrent costs (production, marketing and transaction costs) are assumed to increase by 5% annually. Yields are assumed to remain constant, as well as prices and thus the gross value of the product. The assumption is based on qualitative information gathered in the field. Productivity of French beans has either remained the same or decreased while prices have stagnated over the past five years. The financial CBA analysis results for Kirinyaga, Meru and Nyeri are presented in tables 5.7–5.9. Two scenarios are considered in this analysis: two-season and three-season cropping.

The results presented in Table 5.7 indicate that the only contractual marketing arrangements in Kirinyaga district are profitable under the two-season scenario is the individual contract. As demonstrated earlier, the total yield and prices for individual contracts are considerably higher than other marketing arrangements. Higher yields may be attributed to adherence to prescribed farm and crop management practices, including the application of fertilizers and pesticides. Higher prices are attributed to the improved bargaining position from the producer side (based on quantity and quality of produce) and trust between the farmer and the traders. For the three-season scenario, the NPVs for individual and group contract are positive while the NPVs for scheme contracts are negative. The negative NPVs for scheme contracts and spot market may be attributed to lower prices (Ksh.30 and Ksh.39/Kg of French beans respectively) as well as lower yields compared to individual and group contracts.

Table 5.8 includes the financial CBA of the marketing arrangements in Meru district. In contrast to Kirinyaga, both individual and group contracts are profitable (have positive NPVs) under the two-season scenario. Scheme contracts and spot market on the other hand are not profitable. This indicates that non-compliance with GlobalGAP standards does not necessarily result in higher NPV. The negative NPVs for the two marketing arrangements may be attributed to lower produce prices (Ksh.35 and Ksh.40/Kg of French beans for scheme contracts and spot market exchanges respectively, compared to Ksh.55 and Ksh.50 for individual and group contracts respectively). Moreover, the negative NPV could be associated with the high cost of labour compared to the othe two districts (16% higher than Kirinyaga and about 30% higher than Nyeri).

Farmers and exporters in Meru were observed to have a unique level of cooperation, as well as amongst farmers working in the same group in comparison to Kirinyaga and Nyeri. Producer groups in Meru are more socially cohesive and thus exihibited high levels of trust between farmers and traders. This reduced transaction costs of monitoring and controlling farmers and the need to take precautionary measures, and could partly explain higher prices offered to group contracts in

Table 5.7: Farm level financial cost-benefit analysis of in Kirinyaga district

	Two-season scenario (per acre/year)				**Three-season scenario(per acre/year)**			
	Individual contract	Group contract	Scheme contract	Spot market	Individual contract	Group contract	Scheme contract	Spot market
Sum of Discounted Costs	1,399,122	1,151,436	932,141	1,013,612	1,886,655	1,574,734	1,342,850	1,013,612
Sum of Discounted Benefits	1,402,532	1,078,871	775,439	965,118	2,103,799	1,618,307	1,163,158	965,118
Net present value (NPV)	3,410	(72,565)	(156,702)	(48,495)	217,144	43,573	(179,692)	(48,495)
Benefit-cost ratio	1.00	0.94	0.83	0.95	1.12	1.03	0.87	0.95

Note: ( ) means negative; US$ 1 dollar=Ksh.79

Table 5.8: Farm level financial cost-benefit analysis in Meru district

	Two-season scenario (per acre/year)				Three-season scenario (per acre/year)			
	Individual contract	Group contract	Scheme contract	Spot market	Individual contract	Group contract	Scheme contract	Spot market
Sum of Discounted Costs	1,436,672	1,247,378	986,136	1,108,610	1,949,369	1,701,942	1,422,922	1,108,610
Sum of Discounted Benefits	1,557,620	1,362,075	912,545	997,956	2,336,430	2,043,112	1,368,818	997,956
Net present value (NPV)	120,948	114,697	(73,591)	(110,654)	387,062	341,171	(54,104)	(110,654)
Benefit-cost ratio	1.08	1.09	0.93	0.90	1.20	1.20	0.96	0.90

Note: ( ) means negative; US$1r=Ksh.79

this district compared to the other two districts.[44] The group culture observed in Meru can be described as "hierarchist culture" form of organization according to Paalhaar and Jansen (2011) in their study of group culture and smallholder participation in values chains in Kenya. This is in contrast to the "individualist culture" observed in Kirinyaga, and to some extent in Nyeri. Hierarchical social organization emphasizes strict rules and penalties. This form of organization is guided by high levels of trust and social cohesiveness. On the other hand, individualist cultures are associated with low social cohesiveness and group members with profit-maximization goals (Paalhaar & Jansen 2011). Some of the group characteristics in Kirinyaga, which qualify them as an individualist culture, including low levels of mutual trust and support, a large share of inactive group members and dropouts, a lack of control over side-selling, and leadership wrangles. In contrast, groups in Meru have more active group members (over 95% of total membership), effective mechanisms for controlling side-selling, and strict enforcement of group rules and penalties.[45]

Table 5.9: Farm level financial cost-benefit analysis in Nyeri district

	Two-season scenario (per acre/year)			**Three-season scenario (per acre/year)**		
	Group contract	Scheme contract	Spot market	Group contract	Scheme contract	Spot market
Sum of Discounted Costs	939,539	867,396	978,862	1,278,633	1,243,311	978,862
Sum of Discounted Benefits	1,038,507	747,089	1,019,533	1,557,760	1,120,634	1,019,533
Net present value (NPV)	98,968	(120,307)	40,671	279,127	(122,678)	40,671
Benefit-cost ratio	1.11	0.86	1.04	1.22	0.90	1.04

Note: () means negative; US$1 =Ksh.79

44 Furthermore, Meru district is located further from the Nairobi central district (about 250km away) compared to Kirinyaga and Nyeri districts (approximately 160km) and thus less likely to attract brokers and "briefcase" exporters who encourage side-selling during periods of high demand.

45 This contrast of group culture between in Meru and Kirinyaga was also reflected in the following quotation from a manager of one of the export companies. "*Meru district has the most successful farmer groups. Farmers in Muranga and Kirinyaga district are more difficult to deal with. They are more business minded and therefore likely to divert their produce to any trader who offers them higher prices than the contract price.*" Interview, Senior Manager, Wamu Exporter, 14/05/2012.

Table 5.9 presents mixed results for marketing arrangements in Nyeri district. Group contracts and spot market exchanges are profitable (have positive NPVs) in both the two-season and three-season scenarios. On the other hand, scheme contracts are not profitable, with negative NPV under both production scenarios. The viability of the spot market in Nyeri is in contrast to the results from the other districts, and could be related to lower labour costs (Ksh.170 per man-day). With lower NPVs under scheme contracts in compared to spot market producers, there is limited viability of scheme contracts among the smallholder farmers in Nyeri. This situation may partly explain the abandonment of contract farming by smallholders and from export market vegetables production over the last few years (see Chapter Two). With these results, it can be confirmed that collective action, in the form of producer groups, plays a key role in linking smallholder farmers to markets. As documented in earlier studies (for example Narrod et al., 2009), collective action provides a platform for small producers to negotiate better market terms as also as shown by positive NPVs for group contracts in contrast to non-profitable scheme contracts. Moreover, collective action allows farmers to pool volumes of produce in order to attain economies of scale and to invest collectively in facilities needed to meet production requirements. In addition, trust plays a crucial role to the benefits accrued from collective action. Farmers are likely to gain more in groups that are more socially cohesive and follow strict enforcement of rules and penalties.

5.8.6 The relevance of French beans (export crop) enterprises relative to other farm enterprises

From the results of the financial CBA above, we can conclude that high production costs, including labor costs and compliance with GlobalGAP standards, could be forcing smallholders out of the export business. To ascertain these findings, we examine the change in gross margins for other key farm enterprises that could not only be competing for labour resources, but also generating greater profit in local markets where demand for food crops is increasing with population growth. We examined the changes in gross margin for years 2005 and 2010. This is an important period for agricultural markets. Global food prices increased rapidly, the global financial crisis shocked global markets, and weather instability risks seem to have increased. Dairy farming and the production of Irish potatoes, tomato, dry maize, and green maize are compared to production of the main export crop, French beans. Table 5.10 presents a summary of the gross margins for the above-mentioned enterprises. Kirinyaga district is used for this comparison.

Table 5.10: Gross margin analysis of different farm enterprises

		Year	Value of output (Ksh)	Variable cost (Ksh)	**Gross margin (Ksh)**
Gross margin per acre	Dry maize	_2005	24,000	13,401	10,599
		_2010	40,000	20,558	19,442
	Green maize	_2005	26,667	13,401	13,266
		_2010	48,000	19,558	28,442
	Irish potatoes	_2005	33,000	29,309	3,691
		_2010	49,500	34,055	15,445
	Tomatoes	_2005	82,500	34,225	48,275
		_2010	130,000	37,386	92,614
	French beans	_2005	88,609	54,090	34,520
		_2010	74,165	58,663	15,502
Gross margin per cow	Dairy	_2005	104,718	72,594	32,124
		_2010	163,600	114,374	49,226

Source: Author's data, Ministry of Agriculture (Kirinyaga district) and various secondary sources. US$ 1 dollar=Ksh.79 (nominal values)

Maize and Irish potatoes are typically cultivated twice a year under both rain-fed and irrigated conditions. Tomatoes are also cultivated twice a year, but similar to French beans, they rely heavily on irrigation. The gross margin figure for French beans is the average value across all marketing arrangements discussed in the previous sections. Table 5.10 shows the gross margin trends across all the farm enterprises from 2005 to 2010. French beans yields in 2005 are about 17% higher than in 2010, while the prices remained the same. . In contrast, the prices for other farm enterprises increased in 2010. The gross margin for tomatoes, green maize and dairy increased by about 90%, 100% and 50% respectively. This may be explained by increased demand for those commodities, which have pushed up prices and improved market access. The price of these commodities are estimated to have increased by about 56%, 67%, 43% and 33% for milk, maize, Irish potatoes and tomatoes, respectively, from 2005 to 2010. These domestically consumed products provide a competitive edge to French beans and thus could explain the shift of smallholders to less risky and more certain farm enterprises. For example, the increased number of milk private processors has contributed to rapid growth of the dairy sector. Maize production became a lucrative enterprise since the global food crisis in the year 2008 in combination with erratic weather patterns. A comparison of the gross margin of tomatoes for two plantings to French beans for three plantings reveals that the former is significantly higher than the latter, by about 190%. The National Horticultural

Policy rates tomatoes as the most lucrative vegetable in terms of enterprise value per acre, followed by cabbages and French beans (GoK, 2012).

In addition to lower gross returns, export vegetable crops face high risks compared to domestically consumed vegetables. Over the last few years, farmers and exporters have lost revenues because of stringent requirements in international markets. For example in 2012, US$238 million (Ksh.20 million) worth of fruit and vegetables, including French beans and passion fruits, were denied entry into the EU market because exceeded the maximum permissible pesticide residue levels (0.02 parts per million of dimethoate) (Sambu, 2012). The increased number of interceptions of Kenyan fresh produce has also led to increased controls on Kenyan peas and beans in pods exported to the EU market as indicated in the Commission Implementing Regulation (EU) No. 1235/2012 of December 19, 2012 (EU, 2012). The export business is also challenged by the dynamics of private standards, which continually change. For example, currently exporters are required to comply with GlobalGAP Version 4 that was recently released (GlobalGAP, 2012). The "improvement" requires extra effort, which implies additional transaction costs of maintaining market position. French beans are also highly susceptible to pests and diseases and thus requires more attention compared to other farm enterprises.

5.8.7 Sensitivity analysis

Sensitivity analysis is applied in the CBA to determine which factors are likely to affect the NPV through the adoption of different vegetable marketing arrangements and hence influence their feasibility. Notably, the setting of two and three seasons scenarios explained in section 5.8.5 can be viewed as important variations that determine some of the settings under which different marketing arrangements are feasible. Corresponding to this, variations tested for sensitivity analysis include the price (thus gross value product), labor costs, costs of compliance with GlobalGAP, and the discount rate. Further, variation is examined for fertilizer cost by applying the market price as opposed to the government-subsidized rate. The analysis of the other variables was performed by varying them by 25% (both reductions and increases). Table 5.11 shows the results for Kirinyaga district, while those of Meru and Nyeri district are presented in Appendix 5B.

The sensitivity analysis results presented in Table 5.11 show that all marketing arrangements are profitable under both two and three-season scenarios when product prices are increased by 25%. Similarly, reducing labour cost by 25% has a positive influence on NPV for group contracts and spot market under both

Table 5.11: Sensitivity analysis (Kirinyaga district)

	Two-season scenario				Three-season scenario			
Variations	Individual contract	Group contract	Scheme contract	Spot market	Individual contract	Group contract	Scheme contract	Spot market
Baseline scenario	3,410	(72,565)	(156,702)	(48,495)	217,144	43,573	(179,692)	(48,495)
Increase price by 25%	354,043	197,153	37,157	192,785	743,093	448,149	111,098	192,785
Reduce labour by 25%	142,873	40,418	(45,462)	73,268	426,338	213,047	(12,831)	73,268
Increase fertilizer price	(70,706)	(109,063)	(185,579)	(84,917)	162,511	(11,174)	(223,006)	(84,917)
Decrease discount rate by 25%	(61,805)	(153,924)	(252,135)	(124,462)	174,801	(35,661)	(307,709)	(124,462)
Decrease GlobalGAP investment cost by 25%	106,243	(9,294)	(131,890)	(32,917)	319,977	106,844	(154,879)	(32,917)

Note: ( ) means negative; 1 dollar=Kshs.79

two and three season scenarios. On the other hand increases of fertilizer price to the market rate, renders the NPV for individual contracts under the two-season scenario negative. A similar trend is observed when the discount rate is reduced by 25%. In contrast, decreasing the costs of compliance with GlobalGAP standards does not have any influence on the base NPVs for the different marketing arrangements. Likewise in Meru and Nyeri, increase in vegetable market price and the reduction of labour costs results in positive NPVs for all of the marketing arrangements (see Appendix 5B), except scheme contracts whose NPV remains negative when labour costs are reduced under a two-season scenario. In contrast to Kirinyaga, increasing fertilizer prices does not influence the NPV for both Meru and Nyeri. Decreasing the costs of compliance with GlobalGAP also does not affect the NPV of all the marketing arrangements. From the above sensitivity analysis, it can be concluded that the major drivers of profitability of French bean marketing arrangements are product prices, labour costs and tradable inputs (fertilizer).

5.9 Conclusions and policy implications

The study shows that the changing market conditions are shaping the relationships between traders and farmers with regard to marketing arrangements designs. The external factors driving the commercialization of vegetables result in complex transactions that influence the governance structures of these marketing arrangements. For instance, the introduction of GlobalGAP compliance, defined different contractual arrangements between farmers and exporters. The three main contractual arrangements identified were- individual farmer contracts, group contract and scheme contracts. In addition, some farmers who are not involved in such contract engage in spot marketing. Gross margin analysis of French bean production shows that profitability varies significantly across different marketing arrangements and from one region to another. This could be attributed to the cost of resources such as labour and market influence. Individual contractual arrangements seem to be more rewarding than other marketing arrangements. On the other hand, scheme contracts provide the lowest returns; much lower than spot market producers who rely on market speculation. This reveals that the complexity of the marketing arrangement is positively correlated with profitability.

Using a dynamic financial cost-benefit analysis, this study investigated the feasibility of different marketing arrangements for French beans production. The analysis considered two and three seasons cropping scenarios. Similar to gross margin analysis, the CBA analysis results show a significant variability in feasibility of the different marketing arrangements. In Kirinyaga, individual contracts are financially feasible in both two and three planting seasons. Group contracts are profitable under three planting seasons but not in the two-season scenario. In Meru, individual and group contracts are financially viable under two and three seasons of production. Similarly to Kirinyaga, scheme contracts and spot market exchanges are not profitable in Meru. The results for Nyeri reveal that group contracts and spot market exchanges are feasible under either the two or the three season scenarios, whereas scheme contracts are not financially viable. The profitability of individual contracts across the three districts is evidently due to high yields and higher prices because of better crop management and higher levels of trust between the producer and traders compared to the other marketing arrangements. The role of collective action is also demonstrated by positive NPVs for the group contracts in Meru and Nyeri. Beyond collective action, the advantage conferred by social cohesiveness and mutual trust could also explain the feasibility of group contracts.

The study further compared the gross margin for French beans to food crops that are commonly consumed in local markets considering 2005 and 2010 as the points of reference. The results reveal that while the market prices of most of these commodities increased by 2010, the price of French beans stagnated while productivity dropped. This finding could explain the shift of smallholders to the production of domestic market farm products although further research is required to examine this hypothesis. However, the sensitivity analysis applied in the CBA by varying variables identified to influence the profitability of French bean production across the districts revealed product prices are one of the major drivers of French beans marketing arrangements. The analysis varied the variables by 25% (both deductions and increases). Other important factors identified are labour costs and inputs costs. The sensitivity analysis results does not show any effects on NPV when the costs of compliance with GlobalGAP standards are varied, thus contrasting with earlier claims that compliance with the standards could be driving smallholders out of the high-value market.

A number of policy implications can be derived from the results this analysis. There is a need to lobby for higher prices for smallholders. Moreover, the establishment of a sustainable working relationship between farmers and commercial traders based on mutual trust is profound. Recognizing the role of brokers, middlemen and "briefcase" exporters is necessary to not only ensure high quality produce originating from the country but also to reduce the incidences of side-selling. The initiative by HCDA to register all commercial traders including the seasonal traders should be strengthened as it provides a solution to the long-term problem of side-selling that threatens the export market for vegetables from the Kenya. It is also important to strengthen the recent government program of providing subsidized fertilizer to smallholders, but more so explore long-term strategies such as local production of fertilizers. Furthermore, there is a need to explore the costs and benefits of implementing local standards (such as KenyaGAP standards) as opposed to international standards such as GlobalGAP. KenyaGAP has been recognized and benchmarked within international market standards. Diversification into locally consumed products should also be encouraged among smallholders.

In this study, the profitability of marketing arrangements is based on the private costs and benefits for only a single export crop. Subsequent studies should consider other crops, combination of crops, and other costs and benefits beyond private perspective. An economic CBA is required to determine the costs and benefits of these contractual arrangements at community and national levels.

Appendix 5A: Overview of investment costs, revenues, operating costs and gross margins of French beans production for marketing channels in Kenya (per acre, per year)

Table 5A-1: Kirinyaga district

Kirinyaga district	**Two-season scenario**				**Three-season scenario**			
	Individual	Group	Scheme	Spot market	Individual	Group	Scheme	Spot market
Average yield(Kg)	4160	4000	3833	3670	6240	6000	5750	3670
Average price(Ksh/Kg)	50	40	30	39	50	40	30	39
Gross Value product (GVP)	208000	160000	115000	143130	312000	240000	172500	143130
Production costs	102380	88913	84720	100043	153570	133370	127080	100043
Seeds	18000	17100	16200	18000	27000	25650	24300	18000
Fertilizer	8680	8700	6600	8680	13020	13050	9900	8680
NPK /Foliar	3000	3000	3300	3000	4500	4500	4950	3000
DAP	5680	5700	3300	5680	8520	8550	4950	5680
Chemicals	11820	11593	11150	18063	17730	17390	16725	18063
Irrigation petrol	3840	2880	2880	2880	5760	4320	4320	2880
Labour	60040	48640	47890	52420	90060	72960	71835	52420
Spraying	4000	3040	3040	2200	6000	4560	4560	2200
Irrigation	10260	9500	9500	9500	15390	14250	14250	9500
Weeding	10260	9500	9500	9500	15390	14250	14250	9500
Harvesting	24960	18000	17250	22020	37440	27000	25875	22020
Planting	4560	3800	3800	4000	6840	5700	5700	4000
Land preparation	6000	4800	4800	5200	9000	7200	7200	5200
Marketing costs	2563	2203	3687	1095	3845	3305	5530	1095
Transportation cost	1390	1573	3073	875	2085	2360	4610	875
Transporting facilities	1173	630	613	220	1760	945	920	220
Transaction costs								
GlobalGAP investment cost	85575	37962	16725	6700	85575	37962	16725	6700
Buildings and other facilities	66000	24228	10400	3700	66000	24228	10400	3700
Equipment	11575	7234	3325	2000	11575	7234	3325	2000

Kirinyaga district	**Two-season scenario**				**Three-season scenario**			
	Individual	Group	Scheme	Spot market	Individual	Group	Scheme	Spot market
Protective clothes	5000	4000	3000	1000	5000	4000	3000	1000
Record keeping	3000	2500	0	0	3000	2500	0	0
Additional investment costs	1500	4200	1500	1500	1500	4200	1500	1500
Development fees (group members)		2000				2000		
Group registration fees		700				700		
Water permit	1500	1500	1500	1500	1500	1500	1500	1500
Recurrent transaction costs								
GlobalGAP recurring costs/year	30139	20970	7920	5600	30139	20970	7920	5600
Buildings and other facilities	12240	9632	2500	1400	12240	9632	2500	1400
Equipment	6075	2950	1400	1700	6075	2950	1400	1700
Cost of technical services	1200	1200	0	0	1200	1200	0	0
Protective clothes	3300	2388	1900	500	3300	2388	1900	500
Record keeping	2218	1800	0	0	2218	1800	0	0
Auditing and certification costs	606	0	0	0	606	0	0	0
Other recurrent costs	4500	3000	2120	2000	4500	3000	2120	2000
Additional recurrent transaction costs	1122	4877	987	1000	1122.05	4876.6	987.35	1000
Group fees & certificate renewal	0	1250			0	1250		
Activities with trader/contractor	270	555	415	1000	270.15	555.17	415	1000
Land rent grading shed)	0	100			0	100		
Transport costs (*to and from meetings and other activities such as trainings*)	363	1000	235	0	363.45	1000	235.29	0
Incidental costs (food expenses during the meetings & other activities)	488	771	337	0	488.45	771.43	337.06	0
Loan repayment		1200				1200		
Total operating costs	136,204	116,963	97,314	107,738	188,676	162,522	141,517	107,738
Gross margin	71,796	43,037	17,686	35,392	123,324	77,478	30,983	35,392
Labour productivity (GVP/labour cost)	3.5	3.3	2.4	2.7	3.5	3.3	2.4	2.7
Return to investment(GVP/Inv. cost)	2.4	3.8	6.3	17.5	3.6	5.7	9.5	17.5

Table 5A-2: Meru district

	Two-season scenario				Three-season scenario			
	Individual	Group	Scheme	Spot market	Individual	Group	Scheme	Spot market
Average yield(Kg)	4200	4040	3867	3700	6300	6060	5800	3700
Average price(Ksh/Kg)	55	50	35	40	55	50	35	40
Gross Value product (GVP)	231000	202000	135333	148000	346500	303000	203000	148000
Production costs	107700	95550	91500	110780	161550	143325	137250	110780
Seeds	18000	16867	16200	18900	27000	25300	24300	18900
Fertilizer	8680	8680	6600	8680	13020	13020	9900	8680
NPK /Foliar	3000	3000	3300	3000	4500	4500	4950	3000
DAP	5680	5680	3300	5680	8520	8520	4950	5680
Chemicals	11580	11363	10927	17700	17370	17045	16390	17700
Irrigation petrol	3760	2800	2800	2800	5640	4200	4200	2800
Labour	65680	55840	54973	62700	98520	83760	82460	62700
Spraying	4400	3520	3520	4000	6600	5280	5280	4000
Irrigation	11880	11000	11000	12500	17820	16500	16500	12500
Weeding	11880	11000	11000	12500	17820	16500	16500	12500
Harvesting	25200	20200	19333	22200	37800	30300	29000	22200
Planting	5280	4400	4400	5000	7920	6600	6600	5000
Land preparation	7040	5720	5720	6500	10560	8580	8580	6500
Marketing costs	2660	2297	2520	780	3990	3445	3780	780
Transportation cost	1520	1723	1960	390	2280	2585	2940	390
Transporting facilities	1140	573	560	390	1710	860	840	390
Transaction costs								
GlobalGAP investment cost	87287	38721	17060	6714	87287	38721	17060	6714
Buildings and other facilities (total)	67320	24712	10608	3774	67320	24712	10608	3774
Equipments	11807	7379	3392	2040	11807	7379	3392	2040
Protective clothes	5100	4080	3060	900	5100	4080	3060	900
Record keeping	3060	2550	0	0	3060	2550	0	0

	Two-season scenario				Three-season scenario			
	Individual	Group	Scheme	Spot market	Individual	Group	Scheme	Spot market
Additional investment costs	1700	2100	1500	1500	1700	2100	1500	1500
Development fees (group members)								
Group registration fees	200	600			200	600		
Water permit	1500	1500	1500	1500	1500	1500	1500	1500
Recurrent transaction costs								
GlobalGAP recurring costs/year	28350	25079	8350	5400	28350	25079	8350	5400
Buildings and other facilities	12450	9275	3100	1500	12450	9275	3100	1500
Equipments	5100	5150	1250	1500	5100	5150	1250	1500
Cost of technical services/soil analyses	1300	1250	0	0	1300	1250	0	0
Protective clothes	2500	3211	1700	400	2500	3211	1700	400
Record keeping	1900	2193	0	0	1900	2193	0	0
Auditing and certification costs	100	0	0	0	100	0	0	0
Other recurrent costs	5000	4000	2300	2000	5000	4000	2300	2000
Additional recurrent transaction costs	1220	4585	700	1000	1220	4585	700	1000
Group fees & certificate renewal		1250				1250		
Activities with trader/contractor	270	285	200	1000	270	285	200	1000
Land rent grading shed)		100				100		
Transport costs (meetings& other activities)	550	1200	200		550	1200	200	
Incidental costs	400	550	300		400	550	300	
Loan repayment		1200				1200		
Total operating costs	139,930	127,511	103,070	117,960	195,110	176,434	150,080	117,960
Gross margin	91,070	74,489	32,263	30,040	151,390	126,566	52,920	30,040
Labour productivity (GVP/labour cost)	3.5	3.6	2.5	2.4	3.5	3.6	2.5	2.4
Return to investment (GVP/Inv. cost)	2.6	4.9	7.3	18.0	3.9	7.4	10.9	18.0

Table 5A-3: Nyeri district

	Two-season scenarios			**Three-season scenarios**		
	Group	Scheme	Spot market	Group	Scheme	Spot market
Average yield(Kg)	3920	3760	3600	5880	5640	3600
Average price(Ksh/Kg)	40	30	42	40	30	42
Gross Value product (GVP)	156,800	112800	151200	235200	169200	151200
Production costs	85433	81290	96550	128150	121935	96550
Seeds	17100	16200	18000	25650	24300	18000
Fertilizer	8680	6600	8680	13020	9900	8680
NPK /Foliar	3000	3300	3000	4500	4950	3000
DAP	5680	3300	5680	8520	4950	5680
Chemicals	11593	11150	17370	17390	16725	17370
Irrigation petrol	2880	2880	2880	4320	4320	2880
Labour	45180	44460	49620	67770	66690	49620
Spraying	2720	2720	3200	4080	4080	3200
Irrigation	8500	8500	10000	12750	12750	10000
Weeding	8500	8500	10000	12750	12750	10000
Harvesting	17640	16920	18000	26460	25380	18000
Planting	3400	3400	4000	5100	5100	4000
Land preparation	4420	4420	4420	6630	6630	4420
Marketing costs	2530	2563	705	3795	3845	705
Transportation cost	1840	2050	470	2760	3075	470
Transporting facilities	690	513	235	1035	770	235
Transaction costs						
GlobalGAP investment cost	38283	17371	6536	38283	17371	6536
Buildings and other facilities (total)	23743	10192	3626	23743	10192	3626
Equipments	7090	3259	1960	7090	3259	1960
Protective clothes	4900	3920	950	4900	3920	950
Record keeping	2550	0	0	2550	0	0

	Two-season scenarios			Three-season scenarios		
	Group	Scheme	Spot market	Group	Scheme	Spot market
Additional investment costs	5450	1500	1500	5450	1500	1500
Development fees (group members)	3000			3000		
Group registration fees	950			950		
Water permit	1500	1500	1500	1500	1500	1500
Recurrent transaction costs						
GlobalGAP recurring costs/year	21075	8140	5770	21075	8140	5770
Buildings and other facilities (total)	7565	3040	1600	7565	3040	1600
Equipments	4350	1100	1720	4350	1100	1720
Cost of technical services/soil analyses	1200	0	0	1200	0	0
Protective clothes	2480	2000	450	2480	2000	450
Record keeping	2180	0	0	2180	0	0
Auditing and certification costs	100	0	0	100	0	0
Other recurrent costs	3200	2000	2000	3200	2000	2000
Additional recurrent transaction costs	4120	1520	1000	4120	1520	1000
Group fees & certificate renewal	1250			1250		
Activities with trader/contractor	270	520	1000	270	520	1000
Land rent grading shed)	100			100		
Transport costs (meetings & other activities)	1000	1000	0	1000	1000	0
Incidental costs	300		0	300		0
Loan repayment	1200			1200		
Total operating costs	113,158	93,513	104,025	157,140	135,440	104,025
Gross margin	43,642	19,287	47,175	78,060	33,760	47,175
Labour productivity (GVP/labour cost)	4.1	2.5	3.0	4.1	2.5	3.0
Return to investment (GVP/Investment cost)	4.2	6.0	18.8	6.3	9.0	18.8

Labour cost (Nyeri, Kshs.170 per Man-day; Kirninyaga, Ksh.190 per Man-day ; Meru, Kshs.220 per Man-Day)

Appendix 5B: Sensitivity analysis (Meru and Nyeri districts)

	Two-season scenario				Three-season scenario			
	Individual contract	Group contract	Group scheme	Spot market	Individual contract	Group contract	Group scheme	Spot market
Meru district								
Baseline scenario	120,948	114,697	(73,591)	(110,654)	387,062	341,171	(54,104)	(110,654)
Increase price by 25%	510,353	455,216	154,546	138,835	971,169	851,949	288,100	138,835
Reduce labour by 25%	273,512	244,404	54,103	34,988	615,908	535,731	137,437	34,988
Increase fertilizer price	84,526	78,275	(102,467)	(147,076)	332,429	286,537	(97,419)	(147,076)
Decrease discount rate by 25%	83,855	74,536	(151,215)	(208,670)	384,580	329,280	(155,162)	(208,670)
Decrease GlobalGAP investment cost by 25%	220,282	187,804	(47,651)	(95,536)	486,396	414,278	(28,165)	(95,536)
Nyeri district								
Baseline scenario		98,968	(120,307)	40,671		279,127	(122,678)	40,671
Increase price by 25%		358,594	66,465	295,554		668,567	157,481	295,554
Reduce labour by 25%		186,051	(20,650)	155,930		409,752	26,808	155,930
Increase fertilizer price		68,745	(148,172)	4,249		233,793	(164,476)	4,249
Decrease discount rate by 25%		104,401	(192,204)	(10,311)		318,229	(215,705)	(10,311)
Decrease GlobalGAP investment cost by 25%		154,274	(95,398)	56,581		334,433	(97,769)	56,581

Note: ( ) means negative; US$ 1=Kshs.79

Chapter Six

6. Conclusions

6.1 Synopsis

The study was set out to identify vegetable commercialization pathways for smallholder producers, to assess the factors that affect the decision to participate in those pathways, to assess the effects of market participation on rural livelihoods, and to evaluate the institutions that support small producer participation in those pathways. The general theoretical literature on these subjects are not conclusive on several issues within the discourse of smallholder commercialization and especially in the context of emerging challenges in the face of changing economic, environmental, social and political contexts. The globalization of agriculture for example presents farmers with opportunities as well as challenges such as new production and market regulations and other new risks related to global food prices fluctuations and global financial crises among other concerns.

The study sought to address four specific objectives, which are defined and analyzed in separate chapters. These are: (i) to investigate the pathways of smallholder commercialization of horticultural farming and assess the factors that influence their participation along different pathways, (ii) to assess the implications of commercialization on household income and poverty, (iii) to identify gender roles in the commercialization of horticultural farming and their implications for household welfare outcomes, and (iv) to identify and evaluate the performance of institutions and institutional arrangements that promote smallholder commercialization of horticultural farming.

The study addressed these objectives using two-wave data collected from a random sample of small-scale producer households in five vegetable producing districts in Kenya. The selection of the districts and design of the sampling method was performed by ICIPE, which also collected the first dataset in the 2005/2006 cropping season as part of their Economic Impact Assessment in Horticulture project. The five districts were selected from the Central province (Nyeri, Muramg'a and Kirinyaga) and Eastern province (Meru and Makueni), on the basis of the intensity of vegetable commercialization for both export and domestic markets in the districts. Based on Probability Proportional to Size (PPS) sampling method, 21 sub-locations were then randomly selected and a total of 539 vegetable producer households chosen randomly for survey interviews. This sample provided a sampling frame for the second survey from which a sample of 370 households were randomly selected and revisited by the author between June and August 201. In addition to the household-level interviews, the second survey effort included focus group discussions and interviews with key industry participants.

Chapter Two presents a summary and the results of an applied fixed effects model analysis that accounted for self-selection and a comparison with results of a pooled OLS analysis to explore the factors that explain the choice and extent of participation in vegetable commercialization among smallholders. The models are differentiated based on households participating in the export vegetable market, the domestic vegetable market, or participating jointly in both markets. Further, descriptive statistics are used to explore the trends of smallholder participation in the export and domestic markets and a multinomial logit model to examine the dynamics of farmer participation in the vegetable markets. The results indicate a declining trend of smallholder participation in vegetable commercialization. For instance, about 36% of the households that had been exclusively supplying vegetables to the export market in 2005 had exited this market by 2010. Overall, 20% of the households involved in commercial vegetable production in 2005 had dropped out by 2010. Farmers' perceptions demonstrated that they encountered significant challenges related to the high costs of inputs (seeds, fertilizer and labour) and weather risks such as water shortages and drought. Empirical results show that farmers who had exited from the vegetable market had older household heads, which could reflect labor constraint, and fewer agricultural and livestock assets than those who remained in commercial vegetable production. However, market exitors seem to have access to relatively better roads and productive land than those who continue producing vegetables for sale. This finding highlights the importance of physical infrastructure in offering alternative rural livelihoods other than vegetable production for households who are constrained to participate in vegetable markets. Higher vegetable prices and access to credit favoured participation in the export market while membership in farmer groups and the ability to compliance with GlobalGap standards are positively related to the level of commercialization through this market. Participation in domestic vegetable market was positively related to livestock ownership, ownership of a means of information access (radio), and living greater distances from the nearest market town. Indicators of weather shock and risk are also included in the models to ascertain farmers concerns regarding climate change.

Chapter Three presents a summary and the results of an empirical analysis of the impact of the commercialization of vegetable on per adult equivalent income among smallholders. The analysis also involves detailed descriptive statistics that relates ownership of non-land assets to different vegetable commercialization pathways. The empirical analysis involves triangulation of methods to account for endogeneity due to unobserved heterogeneity and self-selection. In general, the importance of vegetable commercialization to household livelihoods seems to have decreased as farmers diversified into food production and other rural livelihoods such as small rural businesses and off-farm employment. Increases in food prices seem to be lowering the opportunity costs of vegetable production and hence motivating farmers shift away from vegetable production. Empirical results show that vegetables sales generated from the export markets have positive and statistically significant positive effects on per capita income even among the poor households. On the other hand, the results provide weak support for the effect of participation in production of vegetables for the domestic market on per AE income and no effect among the poorer households.

The fourth chapter presents a summary and the results of assessment of gender roles in the commercialization of horticultural farming and an evaluation of the implications of participation by women on household well-being. The descriptive analysis results indicate that women are deprived of access to resources including land, water, agricultural credit, production and marketing information, and extension services. Labour responsibility sharing is evident between men and women. However, correlations between labour input and intra-household management of the income from vegetable sales reveal a linear relationship for male household members but not for female members. Women have more control over lower value vegetables, especially those supplied to the domestic market, while men typically control the production of higher value vegetables supplied to the export market. Probit and Tobit models are used to analyze the determinants of women's participation in the commercialization of vegetables. The results show that female management of export vegetable plots is positively related to the area of land cultivated, the share of cultivated land allocated to vegetables, higher precipitation, and longer distances to the nearest market town. The results also show that female access to extension services and training in vegetable production are positively related to the management of export vegetable production by women. Using cross-sectional data and applying a two-stage control function approach that control for endogeneity due to simultaneity bias; participation in vegetable commercialization by women does not seem to have any impacts on per capita income and assets but is positively related to household expenditure on school fees for children and medical services.

The fifth chapter presents the results of an assessment of different contractual arrangements between smallholders and export market traders. These arrangements are individual farmer contacts, group contracts and scheme contracts. These institutional arrangements have shaped by the changing market conditions in regard to governance structures. Using the commercialization of French bean, a major export market vegetable produced in Kenya, a simple gross margin analysis find that profitability of this enterprise varied by contractual arrangement and district, mainly due to the costs of resources such as labour and market influences. Overall, individual contracts are most rewarding, while scheme contracts provide the lowest returns. Using a dynamic cost-benefit analysis (CBA) underpinned by the transaction costs economics theory, the performance of these institutional arrangements is evaluated and also compared with spot marketing. The feasibility of the contractual arrangements differs across the districts and the seasonality of production (based on two-season or three-season scenarios of annual crop production). Similar to the gross margin analysis results, individual contracts are feasible across all districts while the profitability of the scheme contract is blurred. The positive role of collective action as found in the previous studies is confirmed by the positive profitability of group contracts in Meru and Nyeri. The variability in the profitability of group contracts in different districts could be attributed to differences in social cohesiveness and mutual trust, which we observed to be important determinants of the success of group contracts. Sensitivity analysis results revealed vegetable price, and the costs of labour and inputs (seeds and fertilizers) as the major drivers of French bean production profitability. In contrast to past studies, the costs of complying with Globalgap standards is not identified among the key drivers of French beans production profitability.

6.2 Limitations of the study and suggestions for further research

While the study find a declining trend of smallholder participation in the commercialization of vegetables in Kenya, it does not test empirically the implications of the alternative livelihood options that were adopted by those households that had exited the vegetable markets due to data limitations. This would be an interesting inquest for further research especially to test if alternative livelihoods are better off than participation in commercial vegetable farming. In addition to analyzing poverty from income and asset perspectives, other livelihood aspects such as food security should be considered to provide a better picture of the role of commercialization of farm enterprises on the livelihoods of rural households. Further consideration for monetary values of assets should be made to develop asset index as this reflect the capital stock of a household.

Whereas maize variables in the analysis are included to capture comparative advantages between food crops and commercial vegetables, further research is required to establish the effects of national and global food price instability on the commercial production of vegetables among smallholders.

Although we tried to address the endogeneity and self-selection problems in the analysis of the impacts of women's participation in the commercialization of vegetables on household well-being, the use of household panel data should be considered. For instance, panel data could provide a better understanding of the changes in gender roles and their effects household welfare. The analysis could also be extended to other types of crops and supply chains.

The evaluation of the performance of contractual arrangements is limited to private costs and benefits for only one export crop. Further studies should consider other crops and other costs and benefits beyond the private perspective. An economic cost-benefit analysis is required to determine the costs and benefits of these contractual arrangements at community and national levels.

References

Adekunle, A. A., Ellis-Jones, J., Ajibefun, I., Nyikal, R. A., Bangali, S., Fatunbi, O., & Ange, A. (2012). *Agricultural innovation in sub-Saharan Africa: Experiences from multiple-stakeholder approaches.* Accra, Ghana: Forum for Agricultural Research in Africa (FARA). <http://www.fara-africa.org/media/uploads/library/docs/fara_publications/agrl_innovations_in_ssa.pdf> Accessed September 5, 2012.

Afari-Sefa, V. (2007). The dynamics of horticultural export value chains on the livelihood of small farm households in Southern Ghana. *African Journal of Agricultural Research, 2*(9), 435–440.

Agarwal, B. (1997). Bargaining and Gender Relations: Within and Beyond the Household. *Feminist Economics, 3*(1), 1–51.

Aksoy, A. (2005). The Evolution of Agricultural Trade Flows. In *Global agricultural trade and developing countries* (pp. 17–35). Washington, D.C.: The World Bank.

Alderman, H., Chiappori, P.-A., Haddad, L., Hoddinott, J., & Kanbur, R. (1995). Unitary versus collective models of the household: is it time to shift the burden of proof? *The World Bank Research Observer, 10*(1), 1–19.

Alkire, S., & Santos, M. E. (2010). *Acute Multidimensional Poverty: A New Index for Developing Countries* Oxford Poverty & Human Development Initiative (OPHI) Working Paper No. 38. UK: University of Oxford.

Anand, S., & Sen, A. (1994). *Gender inequality in human development: Theories and Measurement* Human Development Report Office Occasional Paper No. 19. New York: United Nations Development Programme.

Asfaw, S. (2008). *Global Agrifood supply Chain, EU Food-Safety Standards and African Small-Scale producers- The case of High-Value Horticultural Export from Kenya.* Published Doctoral Dissertation, Leibniz Universität Hannover, Germany.

Asfaw, S., Mithöfer, D., & Waibel, H. (2009): Investment in compliance with Global GAP standards: does it pay off small-scale producers in Kenya? *Quarterly Journal of International Agriculture 48*(4): 337–362.

Asfaw, S., Mithöfer, D., & Waibel, H. (2010). Agrifood supply chain, private-sector standards, and farmers' health: evidence from Kenya. *Agricultural Economics, 41*(3–4), 251–263.

Ashraf, N., Giné, X., & Karlan, D. (2009). Finding Missing Markets (and a Disturbing Epilogue): Evidence from an Export Crop Adoption and Marketing Intervention in Kenya. *American Journal of Agricultural Economics, 91*(4), 973 –990.

Attanasio, O., & Szekely, M. (1999). *An Asset-Based Approach to the Analysis of Poverty in Latin America*. IDB-OCE Working Paper No. R-376. <http://ssrn.com/abstract=223582> Accessed March 1, 2011.

Barham, B. L., Foltz, J. D., Jackson-Smith, D., & Moon, S. (2004). The Dynamics of Agricultural Biotechnology Adoption: Lessons from series rBST Use in Wisconsin, 1994–2001. *American Journal of Agricultural Economics*, *86*(1), 61–72.

Barham, J., & Chitemi, C. (2009). Collective action initiatives to improve marketing performance: Lessons from farmer groups in Tanzania. *Collective Action for Smallholder Market Access*, *34*(1), 53–59.

Barrett, C. B. (2008). Smallholder market participation: Concepts and evidence from eastern and southern Africa. *Food Policy*, *33*(4), 299–317.

Barrett, C. B., & Dorosh, P. A. (1996). Farmers' Welfare and Changing Food Prices: Nonparametric Evidence from Rice in Madagascar. *American Journal of Agricultural Economics*, *78*(3), 656–669.

Barrientos, S., Dolan, C., & Tallontire, A. (2003). A Gendered Value Chain Approach to Codes of Conduct in African Horticulture. *World Development*, *31*(9), 1511–1526.

Barrientos, S., Kritzinger, A., Opondo, M., & Smith, S. (2005). Gender, Work and Vulnerability in African Horticulture. *IDS Bulletin*, *36*(2), 74–79.

Bijman, J. (2008). Contract farming in developing countries: an overview. Working Paper. Wageningen University, Department of Business Administration.

Birthal, P. S., Joshi, P., & Gulati, A. (2005). *Vertical coordination in high-value commodities*. International Food Policy Research Institute (IFPRI).

Bizoza, A. R., & de Graaff, J. (2012). Financial cost–benefit analysis of bench terraces in Rwanda. *Land Degradation & Development*, *23*(2), 103–115.

Boughton, D., Mather, D., Barrett, C. B., Benfica, R., Abdula, D., Tschirley, D., & Cunguara, B. (2007). Market Participation by Rural Households in a Low-Income Country: An Asset-Based Approach Applied to Mozambique. *Faith and Economics*, *50*, 64–101.

Bronikowski, A., & Webb, C. (1996). Appendix: A critical examination of rainfall variability measures used in behavioral ecology studies. *Behavioral Ecology and Sociobiology*, *39*(1), 27–30.

Carletto, C., Kilic, T., & Kirk, A. (2009). *Long-Term Welfare Impacts of Export Crop Adoption among Guetemalan Smallholders*. Policy Research Working Paper No. 5142. Washington, D.C.: The World Bank.

Carter, M. R., & Katz, E. G. (1992). Separate spheres and the conjugal contract: Understanding the impact of gender-biased development. *Intra-Household Resource Allocation: Policy Issues and Research Methods, 95–111. Conference proceedings, Washington, D.C.: IFPRI & World Bank.* <http://www.ifpri.cgiar.org/sites/default/files/pubs/pubs/jhu/households/intrahhresch06.pdf> Assessed March 21, 2011.

Chiappori, P.-A. (1992). Collective Labor Supply and Welfare. *Journal of Political Economy, 100*(3), 437–467.

Chichilnisky, G. (2008). The Gender Gap. *Review of Development Economics, 12*(4), 828–844.

Cornwell, C., Schmidt, P., & Wyhowski, D. (1992). Simultaneous equations and panel data. *Journal of Econometrics, 51*(1–2), 151–181.

De Janvry, A., Fafchamps, M., & Sadoulet, E. (1991). Peasant Household Behaviour with Missing Markets: Some Paradoxes Explained. *The Economic Journal, 101*(409), 1400–1417.

De Janvry, A., Sadoulet, E., Fafchamps, M., & Raki, M. (1992). Structural adjustment and the peasantry in Morocco: A computable household model. *European Review of Agricultural Economics, 19*(4), 427–453.

Dercon, S. (1996). Risk, Crop Choice and Savings from Tanzania. *Economic Development and Cultural Change, 44*(3), 485–513.

Dercon, S. (2004). Growth and shocks: evidence from rural Ethiopia. *Journal of Development Economics, 74*(2), 309–329.

Deaton, A. (1991). Saving and Liquidity Constraints. *Econometrica, 59*(5), 1221–1248.

Deaton, A. S. (1989). *Saving in Developing Contries: Theory and Review.* Princeton, Woodrow Wilson School-Development Studies.

Dillon, B. M., & Barrett, C. B. (2013). The Impact of World Oil Price Shocks on Maize Prices in East Africa (e-report). <http://cf.fbe.unimelb.edu.au/seminars/app/UploadedDocs/Doc236.pdf>. Accessed August, 8, 2013.

Dolan, C. (2001). The "Good wife": Struggles over Resources in the Kenyan Horticultural Sector. *Journal of Development Studies, 37*(3), 39.

Dolan, C., & Humphrey, J. (2000). Governance and Trade in Fresh Vegetables: The Impact of UK Supermarkets on the African Horticulture Industry. *Journal of Development Studies, 37*(2), 147–176.

Dolan, C., Humphrey, J., & Harris-Pascal, J. (1999). *Horticulture Commodity Chains: The Impact of the UK Market on the African Fresh Vegetable Industry.* IDS Working Paper No. 96. Brighton: University of Sussex, UK.

Dorsey, B. (1999). Agricultural Intensification, Diversification, and Commercial Production among Smallholder Coffee Growers in Central Kenya. *Economic Geography, 75*(2), 178–195.

Doss, C. R. (2001). Is Risk Fully Pooled within the Household? Evidence from Ghana*. *Economic Development and Cultural Change, 50*(1), 101–130.

Duflo, E., & Udry, C. (2004). *Intra-household resource allocation in Cote d'Ivoire: Social norms, separate accounts and consumption choices*. NBER Working paper No. 10498. USA: National Bureau of Economic Research.

Eaton, C., & Shepherd, A. (2001). Contract farming: partnerships for growth. *FAO Agricultural Services Bulletin* 145.

Ellis, F. (2000). The Determinants of Rural Livelihood Diversification in Developing Countries. *Journal of Agricultural Economics*, *51*(2), 289–302.

English, P., Jaffee, S., & Okello, J. (2004). Exporting out of Africa: the Kenya horticulture success story. *Presented at the Scaling Up Poverty Reduction: A Global Learning Process Conference. Shanghai, China. May 25–27, 2004* (Vol. 25). <http://info.worldbank.org/etools/docs/reducingpoverty/case/120/fullcase/Kenya%20Horticulture%20Full%20Case.pdf> Accessed September, 09, 2011.

European Union (EU) (2012). Commission Implementing Regulation (EU) No. 1235/2012. *Official Journal of the European Union*, *350* (44). <http://www.kephis.org/images/stories/pdf_files/regulation_669_amended.pdf> Accessed June, 29, 2013.

Fafchamps, M. (1992). Cash Crop Production, Food Price Volatility, and Rural Market Integration in the Third World. *American Journal of Agricultural Economics*, *74*(1), 90 –99.

Fafchamps, M. (2004). Market institutions in sub-Saharan Africa: Theory and evidence. *Comparative Institutional Analysis (CIA) Series, vol. 3*.

Fafchamps, M., & Gabre-Madhin, E. Z. (2001). *Agricultural markets in Benin and Malawi: The operation and performance of traders*. World Bank Policy Research Working paper No. 2734. Washington DC: World Bank.

Fafchamps, M., Udry, C., & Czukas, K. (1998). Drought and saving in West Africa: are livestock a buffer stock? *Journal of Development Economics*, *55*(2), 273–305.

Falkingham, J., & Baschieri, A. (2009). Gender and Poverty How Misleading is the Unitary Model of Household Resources? An Illustration from Tajikistan. *Global Social Policy*, *9*(1), 43–62.

FAO, IFAD, & ILO. (2010). Gender dimensions of agricultural and rural employment: Differentiated pathways out of poverty, status, trends and gaps (Eds). Rome: FAO.

Farina, E. M. M. Q., & Reardon, T. (2000). Agri-food Grades and Standards in the Extended Mercosur: Their Role in the Changing Agrifood System. *American Journal of Agricultural Economics*, *82*(5), 1170–1176.

Fischer, E., & Qaim, M. (2011). Smallholder Farmers and Collective Action: What Determines the Intensity of Participation? *Proceeding of German*

Development Economics conference, Berlin, 2011, No. 28. <http://hdl.handle.net/10419/48336> Accessed September 28, 2012.

Fischer, E., & Qaim, M. (2012a). Gender, agricultural commercialization, and collective action in Kenya. *Food Security*, *4*(3), 441–453.

Fischer, E., & Qaim, M. (2012b). Linking smallholders to markets: determinants and impacts of farmer collective action in Kenya. *World Development*, *40*(6), 1255–1268.

Fulponi, L. (2006). Private voluntary standards in the food system: The perspective of major food retailers in OECD countries, *31*(1), 1–13.

Gioè, M. (2006). Can horticultural production help African smallholders to escape dependence on export of tropical agricultural commodities. *Crossroads*, *6*(2), 16–65.

Gitau, R., Olwande, J., Mathenge, M., Kirimi, L., & Kamau, M. (2012). *Potential effects of the imposition of value added tax on agricultural inputs and sifted maize meal.* Policy brief No. 8/2012. Nairobi, Kenya: Tegemeo Institute of Agricultural Policy and Development.

Gittinger, J. (1982). Economic Analysis of Agricultural Projects. (2nd ed.). Washington, D.C.: John Hopkins University Press.

GlobalGAP. (2008). Introduction to GlobalGAP protocal for fresh fruits and vegetables. January version: <http://www.eurep.org> Accessed July 5, 2011.

GlobalGAP. (2012). Integrated Farm Assurance for fresh fruit and vegetables. Edition 4.0, August 2012 version. <www.globalgap.prg> Accessed July 11, 2013.

Goetz, S. J. (1992). A Selectivity Model of Household Food Marketing Behavior in Sub-Saharan Africa. *American Journal of Agricultural Economics*, *74*(2), 444–452.

Government of Kenya (GoK) (2007a). Kenya Vision 2030. <http://chet.org.za/manual/media/files/chet_hernana_docs/Kenya/National/Kenya%20Vision%202030_Popular%20version.pdf> Accessed December 24, 2010.

Government of Kenya (GoK) (2010a). Agricultural Sector Development Strategy, 2010–2020. <http://www.ascu.go.ke/DOCS/ASDS%20Final.pdf> Accessed December 24, 2010.

Government of Kenya (GoK) (2007b). *Basic Report on Well-being on Kenya: Based on Kenya Integrated Household Budget Survey 2005/2006.* Nairobi, Kenya: Kenya National Bureau of Statistics, Ministry of Planning and National Development. < http://statistics.knbs.or.ke> Accessed October 30, 2011.

Government of Kenya (GoK) (2010b). The Land Control Act Chapter 302, Revised Edition 2010 (1989). National Council for Law Reporting with Authority of the

Attorney General. <http://www.kenyalaw.org/klr/fileadmin/pdfdownloads/Acts/LandControlActCap302.pdf> Accessed November 12, 2012.
Government of Kenya (GoK) (2011). The Horticultural Crops Development Authority Legal Order, 2011. <http://www.hcda.or.ke/downloads/Legal%20Notice.pdf> Accessed February 9, 2013.
Government of Kenya (GoK) (2012). National Horticulture policy. Republic of Kenya. <http://www.kilimo.go.ke/kilimo_docs/pdf/National_Horticulture_Policy.pdf> Accessed July 15, 2012.
Goletti, F. (2005). *Agricultural Commercialization, value chains and poverty reduction.* Agrifood Consulting International Discussion paper No. 7. Vietnam: Asian Development Bank.
Govereh, J., Jayne, T. S., & Nyoro, J. (1999). *Smallholder commercialization, interlinked markets and food crop productivity: Cross-country evidence in Eastern and Southern Africa.* Department of Agricultural Economics and Department of Economics Working paper No. 29. USA: Michigan State University.
Graffham, A., Karehu, E., & McGregor, J. (2007). Impact of EurepGAP on small-scale vegetable growers in Kenya. Fresh Insights No. 6. London: International Institute for Environment and Development.
Greene, W. (2012). *Econometric Analysis* (7th ed.). Pearson Higher Education. Edinburgh, England.
Guo, H., Jolly, R. W., & Zhu, J. (2005). Contract farming in China: Supply chain or ball and chain? Proceeding of a conference. Beitrag zur Minnesota International Economic Development. 30th April 2005. USA, University Minnesota.
Haddad, L., Hoddinott, J., & Alderman, H. (1994). *Intra-household resource allocation: An overview.* Working Paper, No. 1255.Washington DC: The World Bank.
Hale, A., & Opondo, M. (2005). Humanising the Cut Flower Chain: Confronting the Realities of Flower Production for Workers in Kenya. *Antipode*, *37*(2), 301–323.
Harris, C., Hegarty, P. V., Kherallah, M. X., Mukindia, C. A., Ngige, J. A., Sterns, P. A., & Tatter, J. (2001). *The Impact of standards on the food sector of Kenya.* Institute of Food and Agricultural Standards research report. USA: Michigan State University.
Horticultural Crops Development Authority (HCDA) (2009). *Annual Horticultural Report, Kenya.* Nairobi, Kenya: Horticultural Crops Development Authority. <http://www.hcda.or.ke/tech/> Accessed November 9, 2010.
Heckman, J. J., & Robb, R. (1985). Alternative methods for evaluating the impact of interventions: An overview. *Journal of Econometrics*, *30*(1), 239–267.
Heltberg, R., & Tarp, F. (2002). Agricultural supply response and poverty in Mozambique. *Food Policy*, *27*(2), 103–124.

Henning, C. H. C. A., & Henningsen, A. (2007). Modeling Farm Households' Price Responses in the Presence of Transaction Costs and Heterogeneity in Labor Markets. *American Journal of Agricultural Economics*, *89*(3), 665–681.

Henry, C., Sharma, M., Lapenu, C., & Zeller, M. (2003). Microfinance poverty assessment tool. Technical Tools. Technical Tool Series No.5. Washington DC: The World Bank.

Henson, S., & Jaffee, S. (2008). Understanding Developing Country Strategic Responses to the Enhancement of Food Safety Standards. *World Economy*, *31*(4), 548–568.

Henson, S., Masakure, O., & Boselie, D. (2005). Private food safety and quality standards for fresh produce exporters: The case of Hortico Agrisystems, Zimbabwe. *Food Policy*, *30*(4), 371–384.

Henson, S., & Reardon, T. (2005). Private agri-food standards: Implications for food policy and the agri-food system. *Food Policy*, *30*(2005), 13.

Hertel, T. W., & Rosch, S. D. (2010). Climate Change, Agriculture, and Poverty. *Applied Economic Perspectives and Policy*, *32*(3), 355–385.

Hobbs, J. E., & Young, L. M. (2000). Closer vertical co-ordination in agri-food supply chains: a conceptual framework and some preliminary evidence. *Supply Chain Management: An International Journal*, *5*(3), 131–143.

Holloway, G., Barrett, C. B., & Ehui, S. (2001). *The Double Hurdle Model in the Presence of Fixed Costs*. Applied Economics and Management Working Paper, Cornell University, New York, US.

Humphrey, J. (2006). Policy implications of trends in agribusiness value chains. *The European Journal of development research*, *18*(4), 572–592.

IFPRI. (2012). Women's empowerment in Agriculture Index. IFPRI.

IMF (2012). *Kenya: Poverty Reduction Strategy Paper- Progress Report*. IMF Country Report No. 12/10. Washington, D.C.: International Monetary Fund. <http://www.imf.org/external/pubs/ft/scr/2012/cr1210.pdf> Accessed August 3, 2012.

Irungu, C. (2002). *Outreach Performance of Non- Governmental Development Organisations in Eastern Kenya*. Published Doctoral Dissertation, Göttingen University, Germany

Jaffee, S. (2003). *From challenge to opportunity: Transforming Kenya's fresh vegetable trade in the context of emerging food safety and other standards in Europe.* Agricultural and Rural Development Discussion Paper No. 2. Washington, D.C.: World Bank.

Jaffee, S., Henson S., Sewadadh M., Lizazo M. B., Ignacio I., Van der meer K. and Andiganacio l. (2005): Food Safety and Agricultural Health Standards:

Challenges and Opportunities for Developing Country Exports. Report No. 31207. Washington, D.C: World Bank.

Jaffee, S., & Little, P. (1994). Contract farming in the shadow of competitive markets: the experience of Kenyan horticulture. In *Little, P. and Watt M. (1994) (Eds). Living under contract: contract farming and agrarian transformation in sub-Saharan Africa* (pp. 97–139). *Madison, Wisc., USA: University of Wisconsin Press.*

Jaffee, S., & Morton, J. (1995). The many faces of success: the development of Kenyan horticultural exports. *In Jaffee, S., and Morton J. (1995) (Eds). Marketing Africa's high-value foods: comparative experiences of an emergent private sector (pp. 319–373). Dubuque, Iowa: Kendall/Hunt Publishing Company.*

Jaffee, S., Van der Meer, K., Henson, S., De Haan, C., Sewadeh, M., Ignacio, L., … Lisazo, M. B. (2005). *Food Safety and Agricultural Health Standards: Challenges and Opportunities for Developing Country Exports*. Discussion Paper No. 31207. Washington, D.C.: World Bank.

Jensen, H. H. (2006). Changes in seafood consumer preference patterns and associated changes in risk exposure. *The Oceans and Human Health*, *53*(10–12), 591–598.

Kangai, E., & Mburu, J. (2012). Opportunities and constraints for youth participation in horticultural production. *Proceedings of Young People, Farming and Food conference, 19–21 March 2012, Accra, Ghana: IFPRI.* <http://www.future-agricultures.org/ypff-conference-resources/cat_view/1569-young-people-farming-and-food-conference/1570-conference-papers> Accessed July 15, 2012.

Karuga, S., & Masabayi, M. (2004). *Domestic Market Study for Fresh Fruit and Vegetables in Kenya*. Nairobi: FINTRAC, Horticulture Development Center.

Karugu, M. N. (2012). The effects of alcoholism in society: a case study of Alcohol consumption patterns in Majengo sub location of Nyeri County. Unpublished Doctoral Dissertation, University of Nairobi, Kenya.

Katz, E. G. (1995). Gender and trade within the household: Observations from rural guatemala. *World Development*, *23*(2), 327–342.

Kennedy, E. (1994). Health and Nutrition Effects of Commercialization of Agriculture. In *von Braun, J and Kennedy, E. (1994) (Eds). Agricultural Commercialization, Economic Development and Nutrition (pp. 79–102). Washington, D.C.: IFPRI. Johns Hopkins, University Press.*

Kennedy, E., & Cogill, B. (1987). Income and nutritional effects of the commercialization of agriculture in southwestern Kenya. *Research Report No. 63. Washington: IFPRI.*

Key, N., Sadoulet, E., & de Janvry, A. (2000). Transactions Costs and Agricultural Household Supply Response. *American Journal of Agricultural Economics*,

82(2), 245 –259.

KHCP. (2011). *Kenya's Intra-Africa Trade in Horticulture*. Nairobi: FINTRAC, Horticulture Development Center.

Kimenye, L. (1993). *Economics of smallholder Flower and French Beans production and Marketing* Published Doctoral Dissertation. Michigan State University.

Kimenye, L. (1995). Kenya's experience promoting smallholder production of flowers and vegetables for European markets. *African Rural and Urban Studies*, *2*(2–3), 121–141.

Kirsten, J., & Karaan, M. (2009). The Economics of Institutions: A review. In *Kirsten, J. et al., 2009. Institutional Economics perspectives on African Agricultural Development* (pp. 24–61). Washington, D.C.: IFPRI.

Kirsten, J., & Sartorius, K. (2002). Linking agribusiness and small-scale farmers in developing countries: Is there a new role for contract farming? *Development Southern Africa*, *19*(4), 503–529.

Kirsten, J., Vink, N., Doward, A., & Poulton, C. (2009). *Institutional Economics perspectives on African Agricultural Development*. Washington, D.C.: IFPRI.

Kristjanson, P., Neufeldt, H., Gassner, A., Mango, J., Kyazze, F., Desta, S., … Coe, R. (2012). Are food insecure smallholder households making changes in their farming practices? Evidence from East Africa. *Food Security*, *4*(3), 381–397.

Kyengo, C. (2006). Improving Agricultural Productivity, Health and Sustainability through Technology and Consumer Demand. Proceeding at the Sustainable Agriculture in Africa from Ideas to Action conference, August 16, 2006, Johannesburg, South Africa: Free Market Foundation. <http://www.policynetwork.net/sites/default/files/kyengo-SDN-16august.pdf> Accessed March 6, 2006.

Lenné, J., & Ward, A. F. (2011). An Approach to strengthening Vegetable Value Chains in East Africa: Potential for Spillovers. In *Mithoefer D. & Waibel H. (2011) (Eds): Vegetable Production and Marketing in Africa: Socio-Economic Research (pp. 195–208). London: CABI.*

Little, P. D. (1994). *Living under contract: contract farming and agrarian transformation in sub-Saharan Africa*. Madison, UA: University of Wisconsin Press.

Macher, J. T., & Richman, B. D. (2008). Transaction cost economics: An assessment of empirical research in the social sciences. *Business and Politics*, *10*(1), 1–63.

Maddala, G. S. (1988). *Introduction to Econometrics* (2nd ed.). Engelwood Cliffs, New Jersey: Prentice-Hall, Inc.

Maertens, M., Colen, L., & Swinnen, J. (2008). *Globalization and Poverty in Senegal: A Worst Case Scenario?* LICOS Discussion Paper series No. 217.

Leuven, Belgium: LICOS center for Institutions and Economic Performance.
Maertens, M., & Swinnen, J. (2009). Trade, Standards, and Poverty: Evidence from Senegal. *World Development*, *37*(2009), 161– 178.
Maertens, M., & Swinnen, J. (2010). Are African High-Value Horticulture Supply Chains Bearers of Gender Inequality? *Gender Dimensions of Agricultural and Rural Employment: Differentiated Pathways out of Poverty*, 155–170.
Manser, M., & Brown, M. (1980). Marriage and Household Decision-Making: A Bargaining Analysis. *International Economic Review*, *21*(1), 31–44.
Markelova, H., & Mwangi, E. (2010). Collective Action for Smallholder Market Access: Evidence and Implications for Africa. *Review of Policy Research*, *27*(5), 621–640.
Masakure, O., & Henson, S. (2005). Why do small-scale producers choose to produce under contract? Lessons from nontraditional vegetable exports from Zimbabwe. *World Development*, *33*(10), 1721–1733.
Mburu, J., Borner, J., Hedden-Dunkhorst, B., Mendoza-Escalante, A., & Frohberg, K. (2007). Feasibility of mulching technology as an alternative to slash-and-burn farming in eastern Amazon: A cost-benefit analysis. *Renewable Agriculture and Food Systems*, *22*(2), 125–133.
McCulloch, N., Winters L. A., and Cirera Xavier (2001). *Trade liberalization and poverty: A Handbook*. London, UK: CEPR.
McCulloch, N., and Ota, M., (2002): Export Horticulture and Poverty in Kenya. IDS working paper No. 174, Brighton, UK: Institute of Development Studies, University of Sussex.
McElroy, M. B., & Horney, M. J. (1981). Nash-Bargained Household Decisions: Toward a Generalization of the Theory of Demand. *International Economic Review*, *22*(2), 333–349.
Mighell, R. L., & Jones, L. A. (1963). *Vertical coordination in agriculture*. Economic Research Service, USDA. Washington DC.
Minot, N. (1986). *Contract farming and its effect on small farmers in less developed countries*. Department of Agricultural, Food, and Resource Economics Working paper. Michigan.: Michigan State University.
Minot, N., & Ngigi, M. (2004). *Are Horticultural exports a replicable Success story? Evidence from Kenya and Cote D'ivoire*. MTID Discussion paper No. 120. Washington DC, IFPRI.
Minot, N., & Roy, D. (2007). Impact of high-value agriculture and modern marketing channels on poverty: An analytical framework. Washington DC, IFPRI.
Miriti, L., Miruka, M., & Maina, I. (2011). Gender and the Banana value chain on Imenti South Distict. In *Advancing Agri-practice: Adding Value for Women*.

KARI,Nairobi, Kenya. <http://www.slideshare.net/ICRW/gender-and-the-banana-value-chain-in-imenti-south-district> Accessed October, 20, 2012.

Mishra, K. K. (1991). Coefficient of variation as a measure of relative wetness of different stations in India. *International Journal of Biometeorology*, *34*(4), 217–220.

Mithofer, D., Nang'ole, E., & Asfaw, S. (2008). Smallholder access to the export market: the case of vegetables in Kenya. *Outlook on Agriculture*, *37*(3), 203–211.

Miyata, S., Minot, N., & Hu, D. (2009). Impact of Contract Farming on Income: Linking Small Farmers, Packers, and Supermarkets in China. *Agri-food Industry Transformation and Small Farmers in Developing Countries*, *37*(11), 1781–1790.

Moser, C., & Felton, A. (2007). *The Construction of an Asset Index Measuring Asset Accumulation in Ecuador*. Chronic Poverty Research Centre Working Paper No. 87). Washington, D.C.: World Bank.

Moti J., Berhanu G., and Hoekstra D. (2009). Smallholder commercialization: Processes, determinants and impact. Discussion Paper No. 18. Nairobi, Kenya: Improving Productivity and Market Success (IPMS) of Ethiopian Farmers Project, ILRI (International Livestock Research Institute).

Muendo, M., & Tschirley, D. (2004). *Improving Kenya's Domestic Horticultural Production and Marketing System: Current Competitiveness, Forces of Change, and Challenges for the Future*. Working Paper No. 08C/2004; Vol III. Nairobi, Kenya: Tegemeo Institute of Agricultural Policy and Development.

Muhammad, A., D'Souza, A., & Amponsah, W. (2013). Violence, Instability, and Trade: Evidence from Kenya's Cut Flower Sector. *World Development*, *51*(0), 20–31. Muriithi, B., Mburu, J., & Ngigi, M. (2010a). *Compliance with International Food Safety Standards: Determinants, costs and implications of EurepGap standards on profitability among smallholders' horticultural exporters in Kenya*. Published Msc. dissertation. Egerton University, Kenya.

Muriithi, B., Mburu, J., & Ngigi, M. (2010b). Constraints and Determinants of Compliance With EurepGap Standards: A Case of Smallholder French Bean Exporters in Kirinyaga District, Kenya. *Agribusiness*, *27* (2), 1–12.

Narrod, C., Roy, D., Okello, J., Avendano, B., Rich, K., & Thorat, A. (2009). Public–private partnerships and collective action in high value fruit and vegetable supply chains. *Food Policy*, *34*, 8–15.

Neven, D., Odera, M. M., Reardon, T., & Wang, H. (2009). Kenyan Supermarkets, Emerging Middle-Class Horticultural Farmers, and Employment Impacts on the Rural Poor. *World Development*, *37*(11), 1802–1811.

Neven, D., & Reardon, T. (2004). The Rise of Kenyan Supermarkets and the Evolution of their Horticulture Product Procurement Systems. *Development Policy Review*, *22*(6), 669–699.

Ngugi, I., Gitau, R., & Nyoro, J. (2006). *Kenya Access to high value markets by smallholder farmers of African indigenous vegetables*. London, UK: IIED.
Niemeijer, R., & Hoorweg, J. (1994). Commercialization of Rice and Nutrition: A case of West Kenya. In *von Braun, J and Kennedy, E.(1994) (Eds). Agricultural Commercialization, Economic Development and Nutrition (pp. 264–275). Washington, D.C.: IFPRI. Johns Hopkins University Press.*
Nimenya, N., Henry de Frahan, B., & Ndirima, P.-F. (2009). Compliance with international food safety standards as an outcome of a Nash bargaining process: a case study on Kenyan small scale green beans farms. Proceedings at the 111th EAAE-IAAE Seminar, 26th–27th June 2009, University of Kent, Canterbury, UK. <http://ageconsearch.umn.edu/bitstream/53004/2/116.pdf> Accessed November, 14, 2010.
Njuki, J., Kaaria, S., Chamunorwa, A., & Chiuri, W. (2011). Linking Smallholder Farmers to Markets, Gender and Intra-Household Dynamics: Does the Choice of Commodity Matter & quest. *European Journal of Development Research*, *23*(3), 426–443.
North, D. C. (1990). *Institutions, institutional change and economic performance*. UK: Cambridge university press.
Nyoro, J., & Jayne, T. S. (2005). *Trends in regional agricultural productivity in Kenya. Kenya agricultural marketing and policy analysis project.* Nairobi, Kenya: Tegemeo Institute of Agricultural Policy and Development.
Odero, D. O., Mburu, J., Ackello-Ogutu, C., & Nderitu, J. H. (2013). Competitiveness of smallholder snap bean production in Kirinyaga County, Kenya, *2*(1), 49–65.
Okello, J. (2005). *Compliance with international food safety standards: The case of Green Bean production in Kenya Family farms*. Published Doctoral Dissertation. Michigan State University, USA.
Okello, J., Narrod, C., & Devesh, R. (2009). Smallholder Compliance with International Food Safety Standards is not a Fantasy: Evidence from African Green Bean Producers. In *de Battisti, A.B., J. MacGregor, and A. Graffham. 2009. Standard Bearers: Horticultural Exports and Private Standards in Africa. Iied.* London, UK: IIED.
Okello, J., Narrod, C., & Roy, D. (2007). *Food Safety Requirements in African Green Bean Exports and Their Impact on Small Farmers*. IFPRI Discussion Paper No. 00737. USA: IFPRI.
Okello, J., & Swinton, S. (2007). Compliance with International Food Safety standards in Kenya's Green Bean Industry: Comparison of a Small- and a large-scale Farm Producing for Export. *Applied Economic Perspectives and Policy*, *29*(2), 269–285.

Olwande, J., & Mathenge, M. (2011). *Market participation among poor rural households in Kenya* Working Paper No. 42. Nairobi, Kenya: Tegemeo Institute of Agricultural Policy and Development.

Omamo, S. W. (1998). Transport Costs and Smallholder Cropping Choices: An Application to Siaya District, Kenya. *American Journal of Agricultural Economics*, *80*(1), 116 –123.

Omiti, J., Otieno, D., McCullogh, E., & Nyanamba, T. (2007). Strategies to Promote Market-oriented smallholder Agriculture in Developing countries: A case of Kenya. Proceeding at the Second AAAE Conference, August 20–22,2007, Accra, Ghana. <http://ageconsearch.umn.edu/bitstream/52105/2/Omiti.pdf> Accessed December, 19, 2012.

Osmani, L. N. K. (2007). A breakthrough in women's bargaining power: The impact of microcredit. *Journal of International Development*, *19*(5), 695–716.

Ouma, S. (2008). *Development practice, agri-food standards, and smallholder certification: The elusive quest for GlobalGAP?* London: IIED.

Paalhaar, J., & Jansen, K. (2011). Group culture and smallholder participation in value chains: French beans in Kenya. *Vegetable production and marketing in Africa: Socioeconomic research*, 97–110.

Pingali, P. (2007). Agricultural growth and economic development: a view through the globalization lens. *Agricultural Economics*, *37*(s1), 1–12.

Pingali, P., Khwaja, Y., & Meijer, M. (2005). *Commercializing Small Farms: Reducing Transaction Costs*. Working Paper No. 05–08. p. 42. Rome: FAO.

Pingali, P., & Rosegrant, M. W. (1995). Agricultural commercialization and diversification: processes and policies. *Food Policy*, *20*(3), 171–185.

Posthumus, H., & De Graaff, J. (2005). Cost-benefit analysis of bench terraces, a case study in Peru. *Land Degradation & Development*, *16*(1), 1–11.

Quisumbing, A. R., & de La Brière, B. (2000). *Women's assets and intra-household allocation in rural Bangladesh: Testing measures of bargaining power*. FCND Discussion paper No. 86. Washington D.C. IFPRI

Rao, E. J. O., & Qaim, M. (2011). Supermarkets, Farm Household Income, and Poverty: Insights from Kenya. *World Development*, *39*(5), 784–796.

Reardon, T., & Berdegué, J. A. (2002). The Rapid Rise of Supermarkets in Latin America: Challenges and Opportunities for Development. *Development Policy Review*, *20*(4), 371–388.

Reardon, T., Timmer, P., Barrett, C. B., & Berdegué, J. A. (2003). The Rise of Supermarkets in Africa, Asia, and Latin America. *American Journal of Agricultural Economics*, *85*(5), 1140–1146.

Rutstein, S. O., & Johnson, K. (2004). *DHS Wealth Index* (DHS Comparative Report). Calverton, Maryland USA: ORC Macro.<http://www.measuredhs.com/pubs/pdf/cr6/cr6.pdf> Accessed July 14, 2012.

Sadoulet, E., & de Janvry, A. (1995). *Quantitative development policy analysis* (Vol. 1). Baltimore & London: John Hopkins University Press.

Sadoulet, E., de Janvry, A., & Benjamin, C. (1998). Household Behavior with Imperfect Labor Markets. *Industrial Relations: A Journal of Economy and Society*, *37*(1), 85–108.

Sambu, Z. (2012). EU fruit rejection forces chemical ban. *Business Daily Africa*. Nairobi.

Sartorius, K., & Kirsten, J. (2005). The boundaries of the firm: why do sugar producers outsource sugarcane production? *Management Accounting Research*, *16*(1), 81–99.

Seshan, G. (2005). The Impact of Trade Liberalization on Household Welfare in Vietnam. World Bank Policy Research Working paper No. 3541. Washington D.C.: World Bank.

Shiundu, K. M., & Oniang'o, R. (2007). Marketing African leafy vegetables, challenges and opportunities in the Kenyan context. *African Journal of Food Agriculture Nutrition and Development*, *17*, 4–12.

Simmons, P., Winters, P., & Patrick, I. (2005). An analysis of contract farming in East Java, Bali, and Lombok, Indonesia. *Agricultural Economics*, *33*(s3), 513–525.

Sindi, J. (2008). *Kenya domestic horticulture subsector: What drives commercialization decisions by rural households?* Unpublished Msc. Dissertation. Michigan State University, USA.

Smartt, J., & Haq, N. (2008). *New Crops and Uses: Their role in a rapidly changing world.* UK: Centre for Underutilized Crops. Southampton, UK: University of Southampton.

Strasberg, P., Jayne, T. S., Yamano, T., Nyoro, J., Karanja, D., & Strauss, J. (1999). *Effects of Agricultural Commercialization on Food Crop Input Use and Productivity in Kenya* . Working Paper No. 41. Washington, D.C.: Michigan State University.

Strauss, J. (1984). Marketed Surpluses of Agricultural Households in Sierra Leone. *American Journal of Agricultural Economics*, *66*(3), 321–331.

Suri, T., Tschirley, D., Irungu, C., Gitau, R., & Kariuki, D. (2009). Rural incomes, inequality and poverty dynamics in Kenya.Tegemeo Working paper No. 30. Nairobi: Tegemeo Institute of Agricultural Policy and Development.

Swamy, G. (2004). *The impact of international trade on gender equality*. PREM-notes No. 86. Washington, D.C.: World Bank.

Tallontire, A., Dolan, C., Smith, S., & Barrientos, S. (2005). Reaching the Marginalized? Gender Value Chains and Ethical Trade in African Horticulture. *Development in Practice, 15*(3/4), 559–571.

Tenge, A., & Hella, J. (2005). Financial efficiency of major soil and water conservation measures in West Usambara highlands, Tanzania. *Applied Geography, 25*(4), 348–366.

Temu A. E., & Temu A. A. (2005): High Value Agricultural Products for Smallholder Markers in Sub-Saharan Africa: Trends, Opportunities and Research Priorities. Proceedings at the High Value Agricultural Products Workshop, International Center for Tropical Agriculture, October 3–5, 2005, Cali, Colombia.

Terza, J. V., Basu, A., & Rathouz, P. J. (2008). Two-stage residual inclusion estimation: Addressing endogeneity in health econometric modeling. *Journal of Health Economics, 27*(3), 531–543.

Thornton, P. K., Jones, P. G., Owiyo, T., Kruska, R. L., Herrero, M., Orindi, V., … Bekele, N. (2008). Climate change and poverty in Africa: Mapping hotspots of vulnerability. *African Journal of Agricultural and Resource Economics, 2*(1), 24–44.

Traill, W. B. (2006). The Rapid Rise of Supermarkets? *Development Policy Review, 24*(2), 163–174.

Tran-Nguyen, A.-N., & Zampeti, B. (2004). *Trade and Gender Opportunities and Challenges for Developing Countries*. United Nations.

Tschirley, D. (2007). *Supermarkets and Beyond: Literature Review on Farmer to Market Linkages in Sub-Saharan Africa and Asia*. Working Paper. USA: Michigan State University.

Tschirley, D., & Ayieko, M. (2008). *Assessment of Kenya's domestic horticultural production and marketing systems and lessons for the future*. Working Paper No. WPS/32/2008. Nairobi, Kenya: Tegemeo Institute of Agriculturall Policy and Development.

Tschirley, D., Muendo, M., & Weber, M. (2004). *Improving Kenya's Domestic Horticultural Production and Marketing System: Current Competitiveness, Forces of Change, and Challenges for the Future* . Working Paper No. 08B/2004; Vol II. Nairobi, Kenya: Tegemeo Institute of Agriculturall Policy and Development.

UN. (2012). *The Millennium Development Goals Report*. New York: United Nations.

USAID. (2011). *Kenya's Intra Africa Horticultural Trade Executive Summary*. Nairobi, Kenya: Fintrac Inc. <http://www.hortinews.co.ke/article.php?id=210> Accessed August 15, 2012.

Verhaegen, I., & Van Huylenbroeck, G. (2001). Costs and benefits for farmers participating in innovative marketing channels for quality food products. *Journal of Rural Studies, 17*(4), 443–456.

Virchow, D. (2008): Indigenous vegetables in East Africa: sorted forgotten, revitalized and successful. In Smartt, J.; and N. Haq (Eds) 2008: New Crops and Uses: Their role in a rapidly changing world. (pp. 79–100). Centre for Underutilised Crops. Southampton, UK: University of Southampton.

Von Braun, J., Bouis, H., & Kennedy, E. (1994). Conceptual Framework. In v*on Braun, J and Kennedy, E. (1994) (Eds). Agricultural Commercialization, Economic Development and Nutrition. Johns Hopkins University Press for the International Food Policy Research Institute* (pp. 11–33). Washington, D.C.: IFPRI.

Von Braun, J., & Immink, M. D. (1994). Nontraditional Vegetables crops and food security among Smallholder Farmers in Guatemala. In v*on Braun, J and Kennedy, E. (1994) (Eds). Agricultural Commercialization, Economic Development and Nutrition. Johns Hopkins University Press for the International Food Policy Research Institute* (pp. 11–33). Washington, D.C.: IFPRI.

Von Braun, J., and Kennedy E. (Eds.) (1994): Agricultural Commercialization, Economic Development, and Nutrition. Washington, D.C.: John Hopkins University Press.

Von Braun, J., & Kennedy, E. (1994). *Agricultural Commercialization, Economic Development, and Nutrition*. Washington, D.C.: IFPRI.

Voors, M. J., & D'Haese, M. (2010). Smallholder dairy sheep production and market channel development: An institutional perspective of rural Former Yugoslav Republic of Macedonia. *Journal of dairy science*, *93*(8), 3869–3879.

Warning, M., & Key, N. (2002). The Social Performance and Distributional Consequences of Contract Farming: An Equilibrium Analysis of the Arachide de Bouche Program in Senegal. *World Development*, *30*(2), 255–263.

Wellard, K. (2011). *Pathways to Commercialisation: Supporting Small Farmers* (CADDP Policy Brief No. 2). UK: Futures Agricultures. <http://www.future-agricultures.org> Accessed July 6, 2013.

Williams, A. P., & Funk, C. (2011). A westward extension of the warm pool leads to a westward extension of the Walker circulation, drying eastern Africa. *Climate Dynamics*, *37*(11), 2417–2435.

Williamson, O. E. (1979). Transaction-cost economics: the governance of contractual relations. *Journal of law and economics*, *22*(2), 233–261.

Wold, B. K. (1997). *Supply response in a gender-perspective. The case of structural adjustment in Zambia*. Statistics, Norway. World Bank.

Wooldridge, J. M. (2002). *Econometric analysis of cross section and panel data*. London: Massachusetts Institute of Technology press.

World Bank. (2007). *World Development Report 2008: Agriculture for Develop-*

ment. Washington, D.C.: World Bank.
World Bank. (2012). *World Development Report 2012: Gender Equality and Development*. Washington, D.C.: World Bank.
World Bank. (2013). *Growing Africa: Unlocking the potential of Agribusiness*. Washington, D.C.: World Bank.
World Bank, FAO, & IFAD. (2009). *Gender in Agriculture sourcebook*. Washington, D.C.
Yaro, J. (2013). The perception of and adaptation to climate variability/change in Ghana by small-scale and commercial farmers. *Regional Environmental Change*, 1–14.
Zeller, M., Sharma, M., Henry, C., & Lapenu, C. (2006). An operational method for assessing the poverty outreach performance of development policies and projects: Results of case studies in Africa, Asia, and Latin America. *World Development*, *34*(3), 446–464.

Survey questionnaire

ZEF Bonn

Zentrum für Entwicklungsforschung
Center for Development Research
Universität Bonn

universitätbonn

***Research topic:* Commercialization of Smallholder Horticultural Farming in Rural Kenya: Poverty, Gender and Institutional Arrangements**

CONFIDENTIALITY:

We are conducting a study on commercialization of horticultural farming in Kenya. This forms the basis for PhD research work for Beatrice Muriithi who is a student at Center for Development Research, University of Bonn, Germany. Your participation is voluntary and can refuse to answer any question at any time. The information you provide will be **CONFIDENTIAL** and findings reported as an aggregate along with those of other farmers. Your privacy will be protected to the maximum.

HOUSEHOLD SURVEY QUESTIONNAIRE

ENUMERATOR NAME ________________________ ENUNAME
ENUMERATOR CODE ________________________ ENUCODE

DATE OF INTERVIEW (DD/MM /YY) /___/___/2011 INTERDATE

DISTRICT____________ DIST SUBLOCATION____________SUBLOC
DIVISION____________ DIV VILLAGE_________________VILLAGE
LOCATION____________ LOC

Only the household head or spouse will be interviewed. No replacement of the household should be done.

Are you interviewing the household head or spouse [__] 1=YES 2=NO **INTVIEWHHSP**

RESPONDENT'S NAME________________________________ RESNAME

RESPONDENT'S POSITION IN THE HOUSEHOLD [____] RESPOSITION

1=Head 2=Spouse 3=Other (specify)

START TIME ______________[SRT] END TIME ______________[END]

HHID: [______] HHID MOB NO:________________________

If the household is **not able to participate** in this survey, why not? (code)_____ **INTERVIEW**

0=Interviewed **1**=head and spouse died **2**=Household moved from this area	**4**=refused **3**=household members cannot be located/found	**5**=Family commitments (wedding, burial etc) **6**=sick **7**=too old

JULY-AUGUST 2011
SECTION 1: HOUSEHOLD DEMOGRAPHY & LABOUR AVAILABILITY:

1.1 Provide the following details about the household head

Names		**HHNAME**
SEX [Code: **1**=MALE, **2**=FEMALE]		**HHSEX**
Age (years)		**HHAGE**
Years of farming experience (years)		**HHFARM**
Which year did you settle in this area? E.g. 1985 (*if born here, ask when his father settled here*)		**HHYRSETT**
Marital status (codes) **(b)**		**HHMARITAL**
Ethnic affiliation		**HHETHNIC**
Number of years of schooling (years)		**HHSCHOOL**

Marital status (b) (HHMARITAL)			
1=Single **2**=Monogamously married	**3**=Polygamous married **4**=Divorced	**5**=widowed **6**=separated	**7**=Other (specify)________

1.2 Who is the **farm owner** (legal owner)? [____] (code **(a)** **FAMROWN**

1.3 Who is the **farm manager**? [____] (code **(a)** **FARMMAN**

Codes (a) 1=Head	**4**=head's father	**6**=Son	**8**=Other joint (specify codes):	**10**=public land
2=Spouse	**5**=head's mother	**7**=Daughter	**9**=Hired worker	**11**=Other (specify):____

1.4 Give details **of all household members (including the household head-HHH)** living permanently on the compound and their primary activities and/or occupations (on and off farm): **INCLUDE CHILDREN AND INFANTS**

Name (first name only)	Age (yrs)	Relationship with HHH (code **(b)**	Sex **1 =M, 2 =F**	Primary (main) Activity Occupation (code **(a)**	Secondary activity (if applicable) (code **(a)**	For those contributing to **farm activities** (adults & children) Average time (hrs) spent on commercialized hort. crop per week		Years of schooling
HMName	**HMAge**	**HMRelation**	**HMSex**	**HMPriAct**	**HMSecAct**	**Hours per week**	**HMAvTime (codes)**	**HMSchool**
1								
2								
3								
4								
5								
6								
7								
8								

Activity codes, (a) (HMPriAct)		**Average time spent on commercial horticultural crop (HMAvTime)**	
0 =None **1** =Horticultural production **2**=Cereal production **3**=Livestock production **4**=Artisan (off-farm activities)	**5**=Casual labourer **6**=Civil servant **7**=Employee **8** =In school/college **9** =Pre-school age **10**=Other (specify)	**0=0** hours **1**=Less than 9 hours **2**=9 hours or above but less than 18hours **3**=18hours or above but less than 27 hours	**4**=27 hours or above but less than 36 hours **5**=36hours and above

Codes for relationship with household head (HMRelation)				
1=head 2=spouse; 3=son/daughter;	4=step child; 5=Father/mother; 6=brother/sister;	7=nephew/niece; 8=son/daughter-in-law; 9=grandchild;	10=unrelated; 11=brother/sister-in-law; 12=Father/Mother-in-law:	13=worker 14=other relative (specify)____

1.5 Have you experienced any major problem or shock (e.g. violence, flood, draught, land slide, fire, theft, death, sickness, livestock death or disease out-break etc) for the last 5 years (since 2006)? **[__] 1=YES 2=NO MAJPROB**

1.5.1 In **YES,** state the problem and the relationship of who was affected to the household head

Problem (write down)	**Who was affected (b) WHOAFFECT**
MAJPROBa	**a**
MAJPROBb	**b**

Codes for relationship with household head (WHOAFFECT)				
1=head 2=spouse; 3=son/ daughter;	4=step child; 5=Father/mother; 6=brother/sister;	7=nephew/niece; 8=son/daughter-in-law; 9=grandchild;	10=unrelated; 11=brother/sister-in-law; 12=Father/Mother-in-law:	13=worker 14=All household members 15=other relative (specify)____

1.6 Did you employ labourers in the year 2010? [___] 1=YES 2=NO (*if NO, go to Section 2*) EMPLOB

1.6.1 If ***YES***, list all those you have employed: their first name, sex, their main activities on the farm, the percentage of time spent on commercialized horticultural crop and their wages.

Name (first name)	Employment duration **(in months)**	Sex **1 =M** **2 =F**	Main activities [Code(**a**)]	Average time (hrs) spent on commercialized hort. crop per week (codes) (**c**)	Wage Amount (Kshs) per unit (**b**)		Are lodging and meals also provided? **1=YES, 0=NO**	
					Amount	Unit (code)	Lodging	Meals
LABNam	**LABDura**	**LABSex**	**LABMaiAct**	**LABAvTime**	**LABAmt**	**LABUnit**	**LABLodg**	**LABMeal**

Main activities (a]) (LABMaiAct)		**Wage unit (b) (LABUnit)**	**Average time spent on commercial horticultural crop (c) (LABAvTime)**	
1=Land preparation 2=Planting & weeding commercial hort. crop(s) 3=Weeding 4=Irrigating 5=Spraying	6 =Harvesting 7=Transporting to collection center 8 =All activities related to hort. Farming 9 =Livestock activities 10 =Food crops activities 11=Household chores 12 =Other (specify)__________	1 =week 2 =month 3=Day 4=other (specify)___	0=0 hours 1=Less than 9 hours 2=9 hours or above but less than 18hours	3=18hours or above but less than 27 hrs 4=27 hours or above but less than 36 hrs 5=36hours and above

SECTION 2: OFF-FARM INCOME & BUSINESS ACTIVITIES

2.1 Provide information regarding details of off-farm and business activities by **all working age** household members (*except the salaried employment pensions and remittances*) during the previous 5 years since 2006 (*including the household head*) living permanently on the compound.

Details		Household members 1	2	3	4
Name (first name only)	**OBName**				
Age (years)	**OBAge**				
Sex **(1=M, 2=F)**	**OBSex**				
Any household member who have **worked off** the household land for cash/kind payment? **1=Yes, 2=No**	**OFFfarm**				
If **Yes**, how long have they worked — For the **last 5 yrs** (give period in years)	**OFFYrs**				
If **Yes**, how long have they worked — For the **last 12 months** (give period in months)	**OFFMonth**				
If **Yes**, what kind of work are they involved in? codes **(a)**	**OFFWork**				
If **Yes,** who was the **employer**? Codes **(b)**	**OFFEmply**				
If **Yes**, what was the payment paid ***(Kshs /month)***	**OFFAmnt**				
Any member who owned a business? **1=Yes, 2=No**	**BUSS**				
If **Yes,** what type of business? (codes) **(c)**	**BUSSType**				
Estimated income **per year** earned from the business. **(codes) (f)**	**BUSSIncom**				

(a) Work codes:	**(b) Employer codes:**	**(c) Type of business**	**(f) Income codes**
1=Farm worker 2=Skilled worker (e.g driver, carpentry, masonry) 3=Unskilled worker 4=Food-for-work 5=Others(specify)__	1=Smallholder horticultural farmer 2=Commercial/large-scale horticultural farmer 3=Other type of crops' farmers (large/and small) 4=Government organization 5=NGOs 6=Others(specify) ________	1=Buying and selling of vegetables 2=Buying and selling of other horticultural crops 3=Artisans(e.g. Bicycle repair, etc) 4=Agricultural input supplier 5=Charcoal selling 6=others (specify):____________	1=<1,500 2=1,501–2,500 3=2,501–5,000 4=5,001–10,000 5=10,001–20,000 6=20,001–30,000 7=>30,001

2.2 Did your household receive any kind of remittances (gifts in monetary form of material, food etc), transfers (including pensions) and other sources of income during the last 12 months? **[___] 1=YES 2=NO REMM**

2.2.1 If **YES (*question 2.2*)**, how much were the remittances or transfers or value of the gifts? **[___] REMMAMNT**

1=<1,500	3=2,501–5,000	5=10,001–20,000	7=>30,001
2=1,501–2,500	4=5,001–10,000	6=20,001–30,000	

SECTION 3: HOUSEHOLD ASSETS

3.1 Provide the following information about your housing conditions and access to facilities.

3.1.1 Roofing type? **1=Grass 2=Iron sheet 3=Bricks** [__]**ROOF**

3.1.2 Type of wall? **1=Stone 2=Mud 3=Timber 4=Iron sheets 5=Other specify**:__________ [__] **WALL**

3.1.3 Is it? **1=Owned 2=Rented 3=Other specify:_______** [__] **HSEOWN**

3.1.4 The **MAIN** type of cooking fuel? (codes) (s) [___] [___] **COOKFUEL**

1=Dung/crop residue 2=Collected firewood	3=Purchased firewood 4=Charcoal	5=Paraffin 6=Gas	7=Solar or electricity

3.1.5 What is the main source of domestic water? (codes) [___][___] [___] **DOMWATER**

1=Running river 2=Own well	3=Public (protected) water spring 4=Unprotected spring	5=Bore hole 6=Piped water	7=other source (specify) ________-

3.1.6 What is the distance to the water source (Km) ________________ **DISTWTERSR**

3.2 Does the household have the following assets (provide per unit market value for the items)

Assets	No. owned now	Current Total Value (Kshs)	Who owns (codes)	Asset	No. owned now	Current Total Value (Kshs)	Who owns (codes)
asset	assqty	assvalue	assown	Asset	assqty	assvalue	assown
Farm assets				28=tractor			
1=wheel barrow				29=ploughs for tractor			
2=spray pump				30=harrow/tiller			
3=boom sprayer				31=planter			
4=water pump				32=ridger/ weeder			
5=Sprinkler				33=combine harvesters			
6=Horse pipes				34=generator			
7=Watering canes				35=green house			
8=other irrigation equipment				**Household assets**			
9=stores (chemical/grain store etc)				36=houses (incl. rental houses)			
10=weighing machine				37=radio			
11=grinder				38=TV			
12=power saw				40=telephone/ mobile phones			
13=water tanks				41=solar panels			
14=animal traction plough				42=sewing/knit-ting machine			
15=milking equipment/shed				43=posho mill			
16=zero-grazing units				44=battery (car)			
17=chaff cutter				45=gas cooker			
18=Motorized/ hand thresher				46=bicycle			
19=cattle dip				47=motorcycle			
20=water trough				48=car			
21=poultry houses				49=truck			
22=pig-stys				50=trailer			
23=borehole or well				51=Furniture (total)			
24=dam				52=Refrigerator			
25=pestle and mortar				53=Computer			

Assets	No. owned now	Current Total Value (Kshs)	Who owns (codes)	Asset	No. owned now	Current Total Value (Kshs)	Who owns (codes)
asset	**assqty**	**assvalue**	**assown**	**Asset**	**assqty**	**assvalue**	**assown**
26=Fishing hook				54=Other (specify)			
27=cart							
Who owns codes: 1=Head	2=Spouse 3=Household (all)	4=Head's father 5=Head's mother		6=Son 7=Daughter	8=Other joint (specify codes) 9=Other (specify)___		

SECTION 4: LIVESTOCK OWNERSHIP

4.1 Do you own livestock? **1=Yes 2=No** [____] **LIVESK**

4.2 If **YES**, Provide the information about the type of livestock in the farm at present and per unit value of the animals (per unit resale value or current market value)

Livestock types	Number owned and present in the farm	No. Purchased in the last 12 months	Who own them (codes) **(a)**	Total value (Kshs)
LIVTyp	**LIVQty**	**LIVPur**	**LIVowner**	**LIVvalue**
1=Calves				
2=Bulls				
3=Oxen				
4=Heifers				
5=Cows				
8=Sheep				
9=Goats				
10=Horses				
11=Donkeys				
12=Mules				
13=Chicken/				
14=ducks/ Geese/Turkeys				
15=Pigs				
16=Rabbits				
17=Beehives				
18=Fish				
19=Other (specify)				
Who owns codes(a): 1=Head	2=Spouse 3=Household (all)	4=Head's father 5=Head's mother	6=Son 7=Daughter	8=Other joint (specify codes) 9=Other (specify)

4.3 Provide details regarding livestock sales in the last 12 months

Product	Did you sell this product/ service? **1=Yes 2=No**	Total revenue received last 12 months (Kshs)	Who received the money **codes (d) in 4.1)**	Who manage the money **codes (d) in 4.1)**
LIVpdt	**Livpdtsell**	**LivRev**	**LivRevRec**	**LivRevMan**
Meat				
Hides/skins				
Milk/cream/butter				
Chicken & eggs				
Honey				
Live Animals (specify)				
Transport services				
Traction (oxen) services				
Breeding services (specify animal type)_				
Others (specify)				

SECTION 5: LAND TENURE & LAND USE

5.1 Please provide the following information about the land used by the household in the last 12 months (also include rented land, and fallow/grazing land)

	Total agricultural cultivated land			Own land left fallow	Land given to other family members		Grazing land			Home stead land
	Own land	Gift land	Rented-in		Rented out	Gift	Own	Rented-in	Obtained as gift	
	cultown	**culgift**	**cultrentin**	**ldfallow**	**rentout**	**ladgift**	**grazown**	**grazrentin**	**grazgift**	**homeld**
Acres										
If you rented out land, how much did you earn in the last 12 months? Kshs]										

5.2 Give details about the plots of land cultivated (including the rented in land) for the last 12 months (2010) in Rainy and dry season, permanent crops (for example coffee) to be recorded in the rainy season crop. For a plot that has **been intercropped/mixed cropped**, for example with 2 crops, divide the size of plot by two. Also provide estimated ***total labour time*** in hours per day and ***number of days per months*** allocated to each crop

Season 1)Rainy 2)Dry 3)All	Plot (no.)	Crop code (a)	Area (acres)	Grown for 1=Home use 2=For sale 3=Both give % of each)	Who owns the plot (codes) (b)	Who manages the plot (codes) (b)	Total labour time allocated: Hrs per day	Number of days/ month	Land quality (codes)	Was the land irrigated 1=Yes, 2=No	If yes, percent-age of land irrigated	Did you use fertilizer **1=Yes 2=No**	Did you use manure (any type) **1=Yes 2=No**	Crop output: Quantity	Units (codes) d)	Cash income (Appx)	Market price (per unit)	Who receives the money (b) if sold
season	**plot**	**crop**	**area**	**crpuse**	**pltown**	**pltmang**	**dyhrs**	**days**	**ldqlity**	**irrig**	**perirr**	**fertuse**	**manure**	**crpqnty**	**crpqnty_ unit**	**crpinco**	**mktpr**	**incorec**

Crop code (a)						**Who own/manage (b)**	**Land quality (c)**	**Units (code d)**
Arrow roots =28	Butternuts=39	Godgets=43	Macadamia nuts =29	Pyrethrum=30	Sweet potatoes=22	1=Head	1=Fertile	1=Kgs
African indigenous vegetables=40	Cabbages=6	Green grams=55	Maize=1	Rice=31	Tea=33	2=Spouse	2=Moderately fertile	2=50Kgs bag
Avocado=54	Capscum=47	Flowers=14	Melons=34	Snow peas=3	Tobacco=27	3=Household (all)	3=Infertile	3=90kgs bag
Baby corns=5	Carrots=52	Fodder=37	Miraa=38	Sorghum=21	Tomatoes=8	4=Head's father		4=numbers
Bananas=53	Cassava =25	French beans=2	Napier grass=36	Spinach=10	Wheat=19	5=Head's mother		5=Wheelbarrow
Barley=18	Chick pea=12	Irish potatoes=23	Ndania=50	Sugar snaps=4	Yam=20	6=Son/Daughter		6=gorogoro
Beans=24	Coffee =17	Lettuce=49	Onions=9	Sugar cane=48	(Other specify)________6	7=Other joint (specify codes)		7=debe
Beetroots=41	Cotton=26	Linseed =15	Pigeon peas=46	Sukuma wiki=7		9=Other (specify		8=cart
Black night shade (managu)=11	Cow peas =13	Lintels =16	Pumpkins=35	Sun flower=32				9=bales
brinjals /biriganya=44	Cucumbers=42							10=pickup
Bracoli=51	Garden peas=45							11=bunches
								13=Other(specify

5.2.1 If selling vegetables (*from 5.2*), when did you start growing vegetables for sale (year)__________**VEGYEAR** (*if does not selling vegetables now go to question* ***5.3***)

5.2.2 Considering the quantity of horticultural produce you sold ***question 5.2*** *above*, how does this amount compare to what you produced five years ago and what do you expect in the coming year(s)?

Comparing last 12 months to ……	
Five years ago…… FIVEAGO	**Next year (s) is expected to be……….NEXTYR**
More [____]	More [____]
The same [____]	The same [____]
Less [____]	Less [____]

5.2.3 Give two reasons for your answer above? HORTREAS

Five years go……FVAGOREA	**Next year (s)…… ………..NEXTYREA**

5.2.4 Have you increased the area allocated to the commercialized horticultural crops over the last 5years **1=YES 2=NO** [___] **INCREARE**

5.2.5 If YES (*question 5.2.4*), what best captures what has happened to household food production and purchases (*tick appropriately*) (*consider major food sources such as maize, beans etc*) **FOODPROD**

	1=True	**2=Partly true**	**3=Not true**
a=Reduced food production area			
b=Reduced quantity of food produced			
c=Reduced food consumption			
d=Increased food purchases			
e=Increased food produced per unit area			
f=Other (specify)____________________			

5.3 If you **DO NOT SELL** vegetables, when did you stop growing vegetables for sale? **YEAR**__________

5.3.1 What are the reasons WHY you stopped growing vegetables for sale? Code [____][____][____] **NOVERGSELL**

1=No market	3=low returns	5=Not interested
2=High production costs	4=Small piece of land	9=Other specify________

SECTION 6: MARKET ACCESS

6.1 What is the distance to the nearest main market center from your farm? (Kms) _______**MRTDIST**

6.2 How far is your farm from the nearest market center in walking hours? _________**MRTWLKHRS**

6.3 What is the cost of transport to the main market (Ksh)? _________**MRTTRACOST**

6.4 What is the type of the **MAIN** road from the farm to that main market? [___] **ROADTYPE**

Road type codes (ROADTYPE)	
1=All seasons tarmac,	3=Seasonal dry road,
2=All seasons murrum road	4=other (specify)

6.5 What is the distance to the bean collection center or product collection centre or shed from the farm? (Kms)______________________**COLLCENDIST**

6.6 How far is your farm from the nearest input store (*if different from 6.2*) in walking (hrs)? ______**INPSTRWLKHRS**

6.7 What is the distance from your farm to the **MAIN** town/urban center (Kms)? _______**TOWNDIST**

SECTION 7: MARKETING AND CONTRACT FARMING

7.1 Provide the following information about buyer of horticultural crops and distribution of income

Type of vegetable produced during the last 12 months **(a)**	*Which market* **1**=Export **2**=Domestic market **3**=1&2	Buyer of the produce (if more than 1, give % for each) Code **(b)**	How long have you been selling to this buyer? **years**	Why did you choose this buyer? Codes **(c)**	Price per unit harvest (as given in **5.2**)	Who receives money from sale **(d)**	Management of income from Vegetables sales		
							% by Man	% by woman	% both
Vegtype	**Mrktype**	**prdbuyr**	**Buyryrs**	**buyrreas**	**prcekg**	**mnyrec**	**perman**	**perwmn**	**perboth**

Name of the vegetables (a)		**Buyers (b)**	**Reasons for choosing main buyer(c)**	**Who owns (d)**
1=French beans 2=Snow peas 3=Sugar snaps 4=Baby corns 5=Tomatoes 6=Cabbages 7=Sukuma wiki 8=Spinach	9=Pumpkins 10=Butter nuts 11=Carrots 12=Irish Potatoes 13=Others (specify) __________	1=Local consumers 2=Large trader 3=Small traders 4=Exporter/agents of exporters 5=Supermarkets 6=middleman/Broker 7=Hotels 8=Institutions (specify)__ 9=Other (specify) ______	1=Nearness to farm 2=Higher prices 3=Stable and reliable prices season after season 4=Minimum rejects per transaction 5=Good reputation 6=Availability 7=(other, specify) ______	1=Head 2=Spouse 3=Household(all) 4=Head's father 5=Head's mother 6=Son 7=Daughter 8=Other joint (specify codes)_______ 9=Other (specify_

7.2 Are you producing under a contract? **1=YES** 2=**NO** [___] **CONTRACT** (*if No, go to question 7.4*)

7.3 If **YES**, (***question*** *7.2*), provide the following information on contractual arrangements with the produce buyer

Type of vegetable produced during the last 12 months **(a)**	Contracted by which buyer (codes) **(b)**	How long have you been contracted (yrs)	Type of contract: 1=**Formal (written) 2=Informal (verbal) on the following;**					Any contract arrangements on inputs supply **1=Yes 2=No**	If Yes, which inputs (codes) **(c)**
			Quantity supplied	Quality supplied	Price	Type of pesticide used	Delivery date		
convegtype	**conprdbuyr**	**conbuyyrs**	**conqnty**	**Conqlity**	**conprc**	**conpestde**	**condeldte**	**coninptss**	**coninptype**

Name of the vegetables (a)		**Buyers (b)**		**Inputs (c)**	
1.French beans 2.Snow peas 3. Sugar snaps 4. Baby corns 5.Tomatoes 6.Cabbages 7=Sukumawiki	8=Spinach 9=Pumpkins 10=Butter nuts 11=Carrots 12=Irish potatoes 13=Others(specify) ______	1=Local consumers 2=Large trader 3=Small trader 4=Exporter/agents of exporters 5=Supermarkets	6=middleman/Broker 7=Hotels 8=Schools 9=Other (specify) ______	1=Seeds 2=Fertilizer 3=Chemicals 4=Technical advice	5=Services for spraying chemicals 6=Cash credit 7=Other (specify______

7.4 Do you have any **DIRECT** arrangements with **INPUT** supplier (**DIFFERENT from the one with the produce** buyer **given above in 7.3**)? 1=**YES** 2=**NO**) [___] **CONINPSPPLIER** (*if No, go to question 7.5*)

7.4.1 If **YES**, provide the following information regarding the input supplier.

Type of Input	Input supplier (name)	Delivery means **(a)**	No. of times delivered in the last 12 months	Distance to trader (KM)	Transport cost (Kshs) per trip (to & from)	Terms of payment **(b)**
suppinptyp	**suppname**	**Suppdeliv**	**supptimes**	**suppdist**	**supptracost**	**supppaymnt**
1=Seeds						
2=Fertilizer						
3=chemicals						
4=technical advice						
5=Services for spraying chemicals						
6=Cash credit						
7=Other (specify)________						

Delivery means (a)		**Terms of payment (b)**	
1=seller delivers to the farm 2=farmer collects from seller premises	3=Other (specify)________	1=Cash of delivery 2=After sale of produce	3=Other (specify)___

7.4.2 What were the requirements for obtaining a contract with either the produce buyer or input supplier? (rank from most important) (consider **MAIN BUYER**) **CONTREQUIR**

	Producer buyer	**Input supplier**
1=Being a member of a group		
2=Have a specific land size (specify in acres)________		
3=Can produce specific quantity of produce (specify)_________		
4=Is within a certain distance from collection center/market (specify, KM)___		
5=Can produce organic crops		
6=Can produce particular horticultural crops		
7=others (specify)________________________		

7.5 Have you received any training on vegetable production for the last **12** months? **1=YES 2=NO** [___] **VEGTRAINING** (*If NO, go to 7.6*)

7.5.1 If **YES**, provide the following information.

Who provided the training?	How many times did you attend (in last 12 months)	Who attended the training **(b)**	Did you pay? **1=Yes, 2=No** If ***Yes***, how much?	Distance to training place (KM)	Transport cost (Kshs) per trip (to & from)
Trainrec	**traintims**	**trainatte**	**Trainpay**	**traindist**	**Traintrscost**

Who provided training (a)			**Who attended the training (b)**	
1=Government agent 2=Exporter 3=NGO 4=Bank 5=Cooperative	6=Research institute (specify)_____ 7=Farmer group member 8=Input dealers	9=Open farmers days (e.g Hort. fair, show)____	1=Head 2=Spouse 3=Household(all) 4=Head's father 5=Head's mother	6=Son 7=Daughter 8=Other joint (specify codes) 9=Other (specify)__ __

7.6 Which of the following captures your knowledge of prices of vegetables that you sell to different produce buyers?

Type of buyer (codes) **(a)** **[prknbyr]**	**1**=I always know prices in advance	**2**=I sometimes know prices in advance	**3**=I never know the prices in advance	**4**=Others (specify)
Type of buyers (a) (prknbyr] 1=Local consumers	2=Large trader 3=Small trader 4=Exporter/ agents of exporters		5=Supermarkets 6=middleman/Broker	7=Hotels 8=Schools 9=Other (specify)_______

7.7 How often do you negotiate on the price of the produce? (*fill in the table)*

Type of buyer (codes) **(a) [prnegbyr]**	**1**=I always negotiate	**2**=I sometimes negotiate	**3**=I never negotiate	**4**=Others (specify)
Type of buyers (a) [prnegbyr]				
1=Local consumers 2=Large trader	3=Exporter/agents of exporters 4=Supermarkets	5=middleman/ Broker 6=Hotels	7=Schools 8=Other (specify) _______	

7.8 How useful are each of the following ***sources of market prices*** information for your vegetables? (*please tick as appropriate using the scale? **1**=not useful **2**=a little useful **3**=Useful **4**=Very useful*

Information source [infosorce]	**1**=Not useful	**2**=A little useful	**3**=Useful	**4**=very Useful
a=From produce buyers (e.g. exporters, brokers etc)				
b=From input traders/supplier				
c=From government extension agents				
d=From neighbours/friends/relatives				
e=media________				
f=internet____________				
g=other specify_______________				

7.9 Does any of the household own a mobile phone? **1=YES 2=NO** [___] (if YES, go to …..) **PHONEOWN**

7.9.1 If **NO**, do you have access to a mobile phone? **1=YES 2=NO** [___] **PHNEACCESS**

7.9.2 If **YES** (*either in question 7.9 and 7.9.1*), provide the following information regarding the use of the mobile phone

Type of use of mobile phone **[usephone]**	Has anyone in the household used a phone for this purpose during the last 12 months **(1=Yes, 2=No) [usepur]**	Rate the importance for this use of a mobile phone for your household **1=not important 2=important 3=very important [phoneimp]**
1=Staying in touch with family & friends		
2=To gather information about farm produce prices		
3=To locate buyers of the farm products or receive orders from farm product buyers		
4=To assist in running your own non-farm business		
5=To send and/or receive money		
6=To get extension information from the providers		
7=Other (specify)____________		

7.10 Which mode describes how you receive your payments for horticultural produce sold?

Type of buyer (use codes **(a)** in 7.7**) BUYERPAY**	Mode of payment (code **below) PAYMODE**

1=Cash 2=Bank	3=SACCOs	4=Through mobile phone system (e.g. mpesa, Zap,etc 5=Other (specify)____________

7.11 When do you receive your revenue of produce sold? (codes) [__] **REVEREC**

Type of buyer (use codes **(a)** in 7.7) **BUYERREV**	When you received cash (code below) **REVEREC**

1=On spot (ie cash on delivery) 2=On an agreed date	3=Late payment 4=Other (specify)________________

7.11.1 If you receive your **payment late**, how long does it take after the agreed date? code [___] **PAYLATE**

1=1–3days	3=Two weeks	5=A month	7=Never sure when I will be paid
2=A week	4=Three weeks	6=After a season	8=Other (specify)____________

7.12 Did you have any of your produce rejected in the last one year by the buyer due to different reasons? (code) 1=**YES** 2=**NO** [___] **PRDREJECT** (*if NO go to 7.13*)

7.12.1 If **YES**, (***question 7.12***), how often was your produce rejected? Codes [____] **REJECTOFT**

1=Often	2=Sometime	3=Few times

7.12.2 If **YES** (***question 7.12***), fill in the following table (refer to the last normal season).

	What percentage of harvest was rejected? **[perreject]**	Did the buyer return the rejected amount? **1=Yes 0=No [rejectretur]**	What did you do with the rejected produce? (codes)**(a) [rejectuse]**
Season 1 (March-May)			
Season 2 (June-Aug)			
Season 3 (Sep-Nov)			
Season 4 (Dec-Feb)			
What did you do with rejected produce? (a)	2=Given to animals	4=nothing/waste	
1=Consumed by household members	3=Sold to another buyer		

7.12.3 Why was the produce rejected? (code) ? [__][__] **REJTREAS**

1=Poor quality **2**=attacked by pests & diseases	**3**=supplied more than agreement with trader **4**=dirty produce	**5**=Have not been certified with required standards

7.12.4 Provide the following as satisfaction regarding what the buyer told you as the reasons for rejecting your produce? (code) [__] **REJTSAT**

1=Not satisfied	2=A little satisfied	3=Satisfied	4=Very satisfied

7.13 If you had an agreement (either formal or informal), is there a time in the last 12 months when your buyer failed to keep his agreement on price or quantity? (*give codes on spaces*)

	1=YES 2=NO	
Quantity supplied	[___]	**qntagrfail**
Price of produce	[___]	**pricagrfail**

7.13.1 If **YES**, (*question 7.13*), how often (code)

	1=Often 2=Sometimes 3=Few times	
Quantity supplied	[___]	**qntagrfailoft**
Price of produce	[___]	**pricagrfailoft**

7.13.2 What do you **often** do when the produce buyer fails to honor the contract? (use table)

	What you did (codes) (a)	
On quantity supplied	[___][___][___]	**qntagrfaildo**
On price of produce	[___][___][___]	**pricagrfaildo**
1=Look for another buyer 2=Demand payment from the buyer for unsold produce 3=withdraw from the contract	4=nothing 5=Other (specify)__________	

7.14 If you have an agreement (either formal or informal), is there a time in the last 12 months when you **sold all or part of your produce to another buyer**? **1=YES 2=NO [___] SIDESEL**

7.14.1 If **YES**, (*question 7.13.2*), how often (code) [___] **SIDESELOFT**

1=Often	2=Sometimes	3=Few times

7.14.2 If **YES**, (*question 7.14.2*), give two main reasons for selling to another buyer. (codes, starting with the most important) [__][__] [__] **SIDESELREAS**

1=Needed immediate cash 2=The other buyer offered higher price 3=The other buyer offered transport	4=Other buyer offered/paid for harvesting services 5=The other buyer provided farm inputs (specify)_______ 6=Other (specify)_______________

7.14.3 What happens when you **fail to honor** your agreement and sell produce to another buyer? [__] **AGGFAIL**

1=Nothing	4=Other (specify)________
2=The buyer does not buy your produce the rest of the season	5=
3=Pay a fine to the group member/or to buyer	

7.15 If selling to an exporter, which exporters are you producing for? Code [___] [___] [___] **EXPORTER**

1=Homegrown	3=Greenlands	5=Sunripe	7=East Africa Growers	9=Wamu
2=KHE	4=Woni	6=Sacco Fresh	8=Frigoken	10=Other (specify)_

7.16 If selling to the domestic supermarkets, which supermarket do you produce for? [___][___] **SUPERMKT**

1=Uchumi	3=Tuskeys	5=Mathai
2=Nakumatt	4=Naivas	6=Other (specify)________

7.17 If selling to **DOMESTIC MARKET** (supermarket or any other), were you previously producing for export market (in the last ten years) but stopped? 1=**YES** 2=**NO** [____] *(if NO, go to 7.18)* **DOMKTSTOP**

7.17.1 If **YES**, what were the reasons for abandoning export production? ***Rank with the most important first.***

Reasons [domktstprea]	**RANK**	**Reasons**	**RANK**
Many compliance requirements	[___]	Low profitability (losses)	[___]
Low productivity of the crop	[___]	Large amounts of rejects by exporters	[___]
Lack of buyers	[___]	Lack/unavailability of required inputs	[___]
High costs of required inputs	[___]	Other (specify) ________	[___]

7.18 If you produce **ONLY FOR DOMESTIC MARKET** (and have never been involved in export production), what are the major constraints for entering into export market (Rank from most important).

Constraints **[expcontr]**	**Rank**	Constraints **[expcontr]**	
1=Inability to find best price	[___]	7=lack knowledge about export market	[___]
2=Capital	[___]	8=Water shortage	[___]
3=Insect, pests and diseases	[___]	9=small farm size	[___]
4=Inability to obtain contract	[___]	10=lack of labour	[___]
5=Inability to comply with quality standards	[___]	11=other (specify)______	[___]
6=Low level of production/ quantity of produce	[___]		

7.19 Do you plan to continue selling to your current buyer? (code) [___] 1=**YES** 2=**NO** *(if YES, go to 7.20)* **CURRNTBUYER**

7.19.1 If **NO**, what are the three main reasons (*starting with most important*? Code [___][___][___] **BUYEREAS**

1=Many compliance requirements 2=Low productivity of the crop 3=Want to increase production which current buyer cannot handle 4=High costs of required inputs	5=Low profitability (losses) 6=Large amounts of rejects by exporters 7=Lack/unavailability of required inputs 8=Other (specify) ________

7.19.2 If **NO**, what alternatives do you have? (rank) **ALTBUYER**

1=Shift to another market channel (specify)____________________	[___]RANK
2=Shift to another crop enterprise production? (specify type)________	[___]
3=Shift to livestock production (specify type)________________	[___]
4=Set-up a business?	
5=Seek employment? (specify type)__________________	[___]

7.20 How often have you experienced shortages and excess production of vegetables in your enterprise over the last 12 months? (tick appropriately)

	1=Always	***2=Sometimes***	***3=Rarely***	***4=Never***	
Shortages					**shortage**
Excess production					**exsspdu**

7.20.1 What happens with vegetables when you have excess production? (code; *most important first*) [__] [___][___] **EXESSPRDUSE**

1=Consumed by household members at home or neighbours	2=Given to animals
3=Sold at a cheaper price (throw-away price)	4=Nothing/waste
5=Others (specify)____________________	

SECTION 8: EXTENSION SERVICES

8.1 Provide the following information on extension access and services.

8.8.1 Extension contact ***for any other crop apart from vegetables***

Did you receive extension contact for any of the farm crop last year? **1=Yes 2=No**	If yes, who was the provider? (*codes*) **(a)**	What types (s) of services were provided? (*codes*) **(b)**	Who in the household was contacted? (codes) **(c)**
Extcont	**extprod**	**extserv**	**exthhrec**

Extension services provider (a)		**Types of services provided (b)**		**Who was contacted (c)**	
1 =Government **2** =NGO/donor **3**=exporter **4**=Local traders **5**=Input dealers **6**=Farmer group	**7**=Co-operative society **8** =private extension service providers **9**=Other specify	1=Product handling 2=Pest management 3=Soil and water use 4=chemical handling	5=Record keeping 6=Field hygiene 7=others (specify________	1=Head 2=Spouse 3=Household (all) 4=Head's father 5=Head's mother	6=Son 7=Daughter 8=Other joint (specify codes) 9=Other (specify

8.1.2 Extension contact **on vegetable production and marketing**

Type of vegetable for which you received extension contact (**codes**) **(e)**	Provider of the service (***codes) (a) in 7.1.1***)	What type(s) of services were provided? (***codes) (b) in question 7.1.1***)	Who in the household was contacted? (***codes) (c) in 7.1.1***)	Did you pay for the services? **1=Yes 2=No**	If **YES**, how much? Kshs.
Extvegtyp	**extvegserv**	**extvegservtyp**	**extvegserec**	**extvegpay**	**extvegcost**

Type of export vegetable (e) 1=French beans 2=Snow peas	3=Sugar snaps 4=baby corns 5=tomatoes	6=Cabbages 7=Sukumawiki 8=Spinach	9=Pumpkins 10=Butter nuts 11=Carrots	12=Irish potatoes 13=Others(specify)____________

SECTION 9: SOCIAL CAPITAL

9.1 Do you or any member of this household belong to a farming group or any other group? 1=**YES, 2=NO** [____] **GROUP** (*if NO, go to 9.5*)

9.2 **If YES** (*question 9.1*), what type of home groups are the household members involved in and who is registered? (*if does not belong to a PMO go to question 9.5*)

Name of local group, association e.g., bean group, dairy group, water group, church group, ROSCA, coffee group etc. **[groupname]**				Who is registered? (***codes***) **(a) [reghhme]**
Codes (a) 1=Head	2=Spouse 3=Household (all)	4=Head's father 5=Head's mother	6=Son 7=Daughter	8=Other joint (specify codes) 9=Other (specify)______

9.3 If any of the household (*from Question 9.2*) belong to a ***Vegetable Production and marketing organization*** (**PMO**)/group, provide the following information regarding the group and number of meetings/joint activities (*if does not belong to a PMO, go to question 9.5*)

Who is registered? ***(codes)*** **(a)**	Year of joining	Who attend those meetings? **(codes) (a)**	Any training offered to the group in last year (Jan–Dec 2010) **1=Yes, 2=No**	If **YES;** Who attend training offered to this group? **(codes) (a)**
pmoreg	**Pmoyr**	**pmomtng**	**pmotrain**	**pmotrnatt**
Codes (a) 1=Head	2=Spouse 3=Household (all)	4=Head's father 5=Head's mother	6=Son 7=Daughter	8=Other joint (specify codes) 9=Other (specify)___

9.4 If **you belong** to a ***Vegetable Production and marketing organization*** (**PMO**)/farmer group

9.4.1 What were the requirements of joining the group? (*rank from most important*)

Requirement [pmorequire]	**Rank**
1=Have a specific land size (specify in acres)__________	
2=Can produce specific quantity of produce (specify)________	
3=Is within a certain distance from collection center/market (specify, KM)________	
4=Known by other group members	
5=Others (specify)_______________	

9.4.2 Please rate your commitment in the group activities (*provide appropriate code*).

Activity [groupactiv]	1=Not committed	**2=A little** committed	3=Committed	4=Very committed
Honoring contracts with produce buyers				
Attending meetings				
Paying monthly contributions				
Other group activities (specify)				

9.5 If **DO NOT BELONG** to a ***Vegetable Production and marketing organization*** **(PMO)**/farmer group;

9.5.1 Were you a member and then dropped from a group? **1=Yes, 2=No** **[__] PMODROP**

9.5.2 If **YES,** why did you drop out from the group? (code) **[__] DROPREAS**

1=Not able to raise group contributions	6=Could not meet the quality of required produce
2=Not able to attend group meetings	7=Wanted to produce/market vegetables independently
3=Reduced quantities of vegetables produced/marketed	8=No longer interested in group
4=Stopped growing vegetables	9=Others (specify)__________

9.5.3 If **NO,** why have you not joined a group? (codes) **[___][___][___][___] NOGROUP**

1=Not able to raise group contributions	5=Limited land size
2=Not able to attend group meetings	6=Cannot find a group nearby
3=Produces small quantities of vegetables	7=Not interested to join
4=Does not grow vegetables	8=Others (specify)__________

SECTION 10: GLOBALGAP STANDARDS

10.1 Have you heard about GlobalGap requirements? 1=**YES** 2=**NO**) [__] **GLOBALGAP** (*if NO, go to section 11*)

10.2 **If YES,** in which year did you learn about Global GAP certification? (MM/YY)………..**GLBGPYEAR**.

10.3 From whom did you first learn about GlobalGap standards? [__][__][__] (most important sources) **GLBGPLRN**

1=Exporter	4=Neighbour	9=Church
2=Middlemen	5=Friends and relatives	10=Newspaper
3=Government extension agent	6=Radio & TV	11=Other (specify)..............

10.4 Are you GlobalGap compliant? **1=YES 2=NO 3=In process [___] GLBG-PCOMP** (*if NO, go to 10.5, then question 10.6*)

10.4.1 **If YES**, categorize the compliance status of the farmer (codes) [___] **GLBCOMTYPE**

1=Individually fully compliant farmer who is an out grower of exporters 2=Group contract farmer (own facilities, production process and keep their records) 3=Group scheme farmer (exporter owns facilities, keeps record and controls Production)

10.4.2 **IF YES,** when did you start complying? **Year______ COMPYEAR**

10.4.3 For which vegetables are you certified for GlobalGap standards? (code) [___] [___] **GLOBCROP**

Name of the vegetables (a)			
1=French beans	3=Sugar snaps	5=Tomatoes	7=Others(specify) ______
2=Snow peas	4=Baby corns	6=Garden peas	

10.4.4 If under **Group Compliance**, who owns the grading shed and certificate? [___] **[GRPCOMSHED]**

1=group owns both 2=group shed but exporter withheld certificate	3=group owns certificate but exporter the shed (contract) 4=exporter owns both (scheme)	5=Other (specify).............

10.5 Have you had formal training in on Global GAP standards? (code) [___] 1=**YES 2=NO. [GLOBTRAIN]**

10.5.1 If **YES**, (*question 9.5*), who facilitated GlobalGap training? **(codes)** [___] [___] **[GLOBTRAFAC]**

1=Ministry of Agriculture	3=Donor organization(specify)____	5=Other (specify)...............
2=Farmer group	4=Exporter	

10.6 If **NOT COMPLIANT**, had you complied with GlobalGap standards, or started the compliance process and then abandoned or failed to renew the standards' certificate? **1=YES 2=NO [__]**

10.6.1 If **YES** (***question 10.10***), what are the main reasons for **ABANDONMENT? (Codes)**[__][__][__] **[ABANDREAS]**

1=Low product prices 2=Constant changes in required inputs	3=High cost of maintenance 4=high volumes of beans rejected despite compliance	5=Other(specify) ______

10.7 What were the major constraints encountered during compliance process? [__][__][__][__] **[GLBCONTR1]**

1=High costs of inputs 2=Adulterated inputs	3=High certification fees 4=Long wait for auditors

10.8 What are the main challenges/constraints of maintenance of GlobalGap compliance? [__][__][__] **[GLBCONT2]**

1=Low product prices 2=Constant changes in required inputs 3=High certification fees	4=high volumes of beans rejected despite compliance 5=Other(specify) ______

10.9 In your opinion what are the benefits of adopting GlobalGap standards to you? (codes, starting with most important) [____][____][____] **GLOBENFT**

1=Stable incomes 2=Higher income /production 3=Higher production	4=Better/improved hygiene 5=Assured market	6=Better prices 7=Better health practices 8=Others(specify)________

10.9.1 If prices and production volumes **have improved** after complying, provide the following details regarding change in production/ prices over time

	Before complying	After complying
Production per season (Kgs)	**kgsbef**	**kgsaft**
Average prices per Kg	**pribef**	**priaft**
Farm management practices that might have changed **after** complying	Change of pesticides	**[__] [__]1=Yes 2=No**
	Change of fertilizers	**[__] [__]1=Yes 2=No**
	Change of fungicides	
	Change of seeds	**[__] [__]1=Yes 2=No**
	Other (specify)________	**[__] [__]1=Yes 2=No**

10.10 If you are **STILL COMPLIANT** (***YES*** *in question 10.4*), with GlobalGap standards, what costs did you undergo over the last 12 months in maintaining the standards? (*Use the table below to guide you*)

Cost category	***Cost component***	*After how long do you replace, repeat or renovate the item or incur the cost after the first time* ***1=weekly, 2=monthly, 3=annually, 4=others (specify)***		***Cost of each replacement, renovation, repeat, etc (Kshs)***	***Support from external source (Kshs)***	***Who supported this external cost (codes) (a)***
		Activity GLBACT	***Period (codes) GLBPRD***	***ACTCOST***	***COSUPPT***	***SUPWHO***
1.Building and facilities (*e.g. cleaning, fumigating, painting, repairing etc*) **[BUID]**	Storage room for pesticides **[BULD1]**					
	Storage room for fertilizers **[BULD2]**					
	Packing houses **[BULD3]**					
	Toilet facilities **[BULD4]**					
	Irrigation station **[BULD5]**					
	Grading shed **[BULD6]**					
	Cooler **[BULD7]**					
	OTHER__					
2.Equipments (*replacement or repair*) **[EQU]**	For pesticide delivery **[EQU1]**					
	For irrigation **[EQU2]**					
	For water and soil analysis **[EQU3]**					
	Other Equipment **[EQU4]**					

Cost category	*Cost component*	*After how long do you replace, repeat or renovate the item or incur the cost after the first time* ***1=weekly, 2=monthly, 3=annually, 4=others (specify)***		***Cost of each replacement, renovation, repeat, etc (Kshs)***	***Support from external source (Kshs)***	***Who supported this external cost (codes) (a)***
		Activity GLBACT	***Period (codes) GLBPRD***	***ACTCOST***	***COSUPPT***	***SUPWHO***
5.Technical assistance/ services **[TECH]**	Soil and water analysis **[TECH1]**					
	Spraying **[TECH2]**					
6. Input use and protective clothes [INPUSE]	Masks **[INPUSE1]**					
	Gloves **[INPUSE2]**					
	Other protective clothes **[INPUSE3]**					
7.other cost of certification **[OTCOS]**	Annual audit **[OTCOS1]**					
	Certificate renewal **[OTCOS2]**					
	Record keeping **[OTCOS3]**					
	Consultation fee **[OTCO4]**					
	Other costs **[OTCOS5]**					
Who supported this external cost (a) 1=Produce buyer (exporter/supermarket/broker)		2=Ministry of Agriculture/ Government extension agent		3=NGOs	4=Other (specify)_______	

10.11 Have you ever looked for alternative markets with fewer requirements for your vegetables? [__] **1=YES 2=NO** **ALTMRKT**

10.12 If you are **CERTIFIED** with GlobalGap standards OR **HAD COMPLIED AND THEN ABANDONEND COMPLIANCE or HAD RECEIVED FORMAL TRAINING ON GlobalGap standards**; are you applying the same code of practice for other crops produced for home consumption or for domestic markets? (code) **[__] 1=Yes 2=No 3=Partly GLBTHRCROP**

10.12.1 If **YES or Partly,** what are the major **three** practices you apply to other crops? **GLOPRACT**

1=____

2=____

3=____

10.12.2 If **NO**, give **two** major reasons why you don't apply to other crops? **GLOPRACNO**

1=____

2=____

SECTION 11: CREDIT, FINANCING AND PATTERNS OF INTERACTION

11.1 Did you obtain any credit (either in cash or in kind) in the last 12 months? [___] **1=YES 2=NO CREDIT**

11.2 If **YES**, for what purpose? [____] **CREPURP**

1=School fees	3=household needs	5=Buy livestock(specify)_______
2=Medical care	4=To buy farm inputs	6=Other (specify)____________

11.3 Did you receive credit (in cash or kind) for agricultural production in the last 12 months? [___] **1=YES 2=NO (***if NO, go to question 11.5)* **CRDTYPE**

11.4 If **YES**, did you receive any credit on vegetable production and marketing? **1=YES 2=NO [__] VERCRED (*If No, go to 11.5*)**

11.4.1 If YES, please fill in the table below.

Type of vegetables for which you obtained credit (codes) **(a)**	Input type received or bought on credit (codes) **(c)**	**Major form of credit** 1=Cash 2=in Kind 3=Others (specify)_	Quantity received or bought	**Unit of input** 1=grammes 2=Kgs ; **3**=Litres 4=acres ; **5**=hours supported 6=90kg bag 7=others(specify)_	Value input per unit specified	Total cost (Kshs	Major Source of credit **(codes) (b)**	**How was the credit repaid** 1=crop output (in kind) 2=Livestock revenue 3=off-farm income 4=crop revenues 5=2& 4	Who in the household received the credit **(codes) (d)**
Crevegtyp	**creinpt**	**creform**	**credqnty**	**credunit**	**creuntval**	**cretotct**	**cresour**	**crerepay**	**credhhrec**

Type of the vegetables (a)		**Major source of credit (b)**		**Input type codes (c)**		**Codes (d)**
1=French beans 2=Snow peas 3=Sugar snaps 4.=Baby corns 5=Tomatoes 6=Cabbages 7=Sukumawiki 8=Spinach	9=Pumpkins 10=Butter nuts 11=Carrots 12=Irish potatoes 13=Others (specify)____	**1**=Government fund/ agency e.g AFC **2**=exporters/buyers **3**=Commercial bank **4**=Informal lenders **5**=Donor/NGO/MFI's	**6**=groups (farmer groups, ROSCAS) **7**=Relatives/friends **8**=Co-operative/SACCO **9** =Other Specify	**1**=Fertilizer 2=Seeds 3=Pesticides 4=Insecticide 5=Herbicide 6=Sprayer 7=plough 8=Fungicide	9=Land preparation 10=Technical support 11=Other farming activities ________ 12=Irrigation equipment 13=Crates 14=Cash 15=others(specify)__	**1**=Head 2=Spouse **3**=Household (all) 4=Head's father 5=Head's mother 6=Son 7=Daughter 8=other (specify)_

11.5 If **NO** (*question 11.4*), why haven't you obtained credit? (Codes)
[__] [__] [__] **CREDNO**

1=Does not need a loan 2=Does not have required collateral	3=Not a member of the (Microfinance institution (MFI)	4=High cost of obtaining the loan/credit 5=Other (specify)__________

SECTION 12: HOUSEHOLD DISTRIBUTION OF INCOME, CONSUMPTION AND WEALTH

12.1 Please indicate your **monthly** household expenses (Kshs)

Food **Ksh**_____	School fees **Ksh**______	Entertainment **Ksh**_____
Clothing **Ksh**_____	Medicare **Ksh**______	

12.2 Rank the different sources of income to the household and provide **ANNUAL** estimate by source. **For ranking: 1=Main source of income, 2=2nd source 3=3rd source etc.**

Source of income		**Rank**	Annual estimate (Kshs)	Income managed by both adult male & female (%)	Income managed by adult male (%)	Income managed by adult females (%)
Income from exported horticultural crop	**expinco**					
Income from locally sold horticultural crops	**dominc**					
Income from other farm crops	**cropinc**					
Income from livestock sales and livestock products (e.g. milk)	**livinc**					
Income from other farm activities (e.g. brew making, charcoal burning etc), other specify……………	**fmactinc**					
Income from wages/salaries/ non-farm, pension and (specify profession)_	**wagsinc**					
Income from business activities	**businc**					
Income from remittances/gifts from absent family members and other external income	**reminc**					
Income from rental houses	**rentinc**					
Income from other sources, specify:	**othrinc**					

12.3 Do you have any of the following insurance for yourself, your family, crops and livestock (code)

Type of insurance	1=Yes 2=No	Type of insurance	1=Yes 2=No
1=Private health insurance **INS1**		4=Livestock insurance **INS4**	
2=Public health insurance (NHIF) **INS2**		5=Other (specify)______ **INS5**	
3=Crop insurance (specify crop)__ **INS3**			

SECTION 13: DIETARY DIVERSITY INDICATORS

13.1 Please provide the following information about all the different foods that you have eaten in the last 30 days. Tell us whether you ate the following foods

J=16 to 30 days in the last days (at least every day) **S**=4–15 days in the last one month (once or twice in a week)	**M**=1–3 days in the last month **R**=0 days in the last one month

Food item (FOOD)	Frequency (FREQ)				Food item (FOOD)	Frequency (FREQ)			
	J	S	M	R		J	S	M	R
Cereals					**Fruits**				
1=Maize					21=Bananas				
2=Rice					22=Oranges				
3=Millet					23=Pawpaws				
4=Sorghum					24=Mangoes				
5=Bread /Chapati					25=Pineapple				
6=Other cereals (specify)					26=Lemons				
Roots and Tubers					27=Avocado				
					43=Other fruits				
7=Irish potatoes					**Meat**				
8=Sweet potatoes					28=Beef				
9=Cassava					29=Goat /sheep				
10=Ground nuts					30=Chicken				
11=Other tubers					31=Fish (any)				
Vegetables					32=Other sea food				
12=Sukuma wiki					33=Other meat(specify				
13=French beans					**Milk products**				
14=Spinach					34=Cow milk				
15=Tomatoes					35=Goat milk				
16=Onions					36=Butter				
17=Carrots					37=Other milk products				
18=Okra					**Other items**				
19=Other vegetables					38=Beans				
20=African indigenous vegetables					39=Eggs				
					40=Edible oils				
					41=Sugar				
					42=Honey				

13.2 MAIZE STOCKS

13.2.1 How many 90 kg bags of **maize** did you have **in stock** from **your own production** just **before** you began **harvesting** your 2010/2011 main season maize crop (Jan-March 2011)_____ (bags) **MZESTCKBE**

13.2.2 How many 90 kg bags of **maize** do you have **in stock** right **now from the last harvest?** __________ (bags) **(record "=0" if the household did not plant maize.)** **MZESTCKNOW**

13.2.3 **IF** ***question 13.2.2=0*** (no maize stocks), in which **month** and **year** did you run out of maize stocks from your own production? __________ (month) ______(Year)

1=January 2=February	3=March 4=April	5=May 6=June	7=July 8=Aug	9=September 10=October	11=November 12=December

13.3 Did you receive relief food for the last 12 months? **1=Yes 2=No [__] RELIEF**

13.3.1 If **YES**, how many months______ **RELIEFMNT**

END

THANK YOU

(Please remember to thank the farmer genuinely)

The enumerator to answer section 14 below privately immediately after the interview.

14.1 In your opinion, how did you establish rapport with this respondent [__] **RAPP**

1=with ease	2=with some persuasion	3=with difficulty	4=it was impossible

14.2 Overall, how did the respondent give answers to the questions [__] **RESPOVER**

1=willingly	2=reluctantly	3=with persuasion	4=it was hard to get answers

14.3 How often do you think the respondent was telling the truth [__] **REPTRTH**

1=rarely	2=sometimes	3=most of the times	4=all the time

I (the enumerator) certify that I have checked the questionnaire two times to be sure that all the questions have been answered, and that the answers are legible.

Signed: ____________________________________ Date___/___/_____

Abstract

The horticultural sub-sector in Kenya like in other Sub-Saharan countries is known for its rural growth and pro-poor impacts due to its contributions to income and employment generation and to food security. Moreover, these impacts are expected to be enhanced with the growing demand for the commercialization of horticultural farming as a result of globalization, urbanization and population growth. However, there are concerns that despite the prevailing opportunities, smallholder participation is declining due to challenges such as climate risks, stringent food-safety regulations imposed in the global market supply chains, and other institutional challenges. This study, which involves two rounds of rural household survey in 2005 and 2010, conducted in five major vegetable producing districts in Kenya, seeks to access income and poverty effects of vegetable commercialization, with a special focus on gender issues. In addition to analyzing the dynamics of smallholders participation in vegetable commercialization, the study also identifies and evaluates the performance of institutional arrangements that link the small producers to the high-value vegetable supply chains.

The participation of smallholder in the domestic and export vegetable markets is indeed declining. For instance, 36% of those who participated in the export trade in 2005 had exited by 2010. The declining trend is associated with weather risks and high costs of inputs and unskilled labor, all of which doubled during this period, as well as erratic vegetable prices especially in the international markets. Using a multinomial logit model and a two-stage panel data model, the findings shows that access to credit and vegetable prices are the major factors that influence participation in the export vegetable markets while ownership of non-land assets including information access facilities positively influence participation in the domestic vegetable markets. The analysis of the impact of vegetable commercialization on per capita income confirms that poor households may benefit from vegetable commercialization. A two-step fixed effect model that accounts for both observed and unobserved heterogeneity estimates that households supplying vegetables to the export market have per capita income approximately 39% higher compared to non-participants of this market. The examination of gender roles in vegetable commercialization shows that women are deprived of access to land, water, agricultural credit and production and marketing information. The empirical analysis however establishes that improvement of land productivity and promoting women's access to agricultural training and extension services might enhance their participation in vegetable commercialization. The evaluation of the contractual arrangements between small vegetable producers and exporters reveals that the governance structure of different arrangements has a strong influence on the transactions between these parties, and these affect the profitability of the arrangements, and may vary from one region to another. In addition, a dynamic financial cost-benefit analysis of French bean farming finds that individual contractual arrangements are more profitable than group contracts, scheme contracts and spot marketing.

Key policy interventions that may enhance vegetable commercialization among smallholders include increasing market access through the improvement of physical infrastructure and the provision of low-cost rural credit to facilitate the acquisition of farm inputs such as seeds and fertilizers. In addition, gender training should be incorporated in designing support for market value chains. Further, collective action and contractual arrangements between small producers and traders should be strengthened, including recognizing the role of brokers, intermediaries and seasonal exporters

Zusammenfassung

Dank seinem Beitrag zur Generierung von Einkommen und Arbeitsplätzen ist der Gartenbau in Kenia und anderen Ländern südlich der Sahara bekannt für Wirtschaftswachstum in ländlichen Gegenden und seine armutsmindernde Wirkung. Es is zu erwarten, dass diese Effekte infolge steigender Nachfrage durch Globalisierung, Urbanisierung und Bevölkerungswachstum weiter zunehmen werden. Es gibt jedoch Bedenken, dass die Beteiligung von Kleinbauern trotz dieser vielfältiger Möglichkeiten aufgrund Herausforderungen wie dem Klimawandel, strengen Nahrungssicherheitsvorschriften globaler Wertschöpfungsketten oder anderen institutioneller Rahmenbedingungen abnimmt Die vorliegende Studie beruht auf zwei Runden einer Haushaltsbefragung im ländlichen Kenia, die 2005 und 2010 in fünf verschiedenen gemüseproduzierenden Gegenden durchgeführt wurden. Die Studie untersucht die Auswirkungen der Kommerzialisierung des Gemüseanbaus auf Einkommen und Armut unter besonderer Berücksichtigung der Geschlechterfrage. Außer der Dynamik in der Beteiligung von Kleinbauern bei der Kommerzialisierung des Gemüseanbaus identifiziert und evaluiert die vorliegende Studie auch die Auswirkungen verschiedener institutioneller Arrangements auf die Verbindung von Kleinbauern zu hochpreisigen Gemüse-Wertschöpfungsketten.

Die Beteiligung von Kleinbauern im Binnen- und Exportmarkt für Gemüse geht zurück. Zum Beispiel haben 36% der Kleinbauern, die 2005 am Exportmarkt teilgenommen haben, den Markt bis 2010 verlassen. Der absteigende Trend hängt einerseits mit Wetterrisiken und hohen Kosten für Inputs und ungelernte Arbeiter zusammen, die sich in dieser Zeitspanne verdoppelt haben, und stark fallenden Preisen für Gemüse insbesondere auf internationalen Märkten andererseits. Mit Hilfe eines multinomialen Logit-Modells und einem zweistufigen Paneldaten-Modell wird gezeigt, dass Zugang zu Krediten und Gemüsepreise die wichtigsten Faktoren sind, die die Beteiligung von Kleinbauern am Exportmarkt beeinflussen. Eigentum von Vermögen (außer Land), einschließlich der Möglichkeiten zur Kommunikation, beeinflussen die Beteiligung am Binnenmarkt positiv. Die Analyse des Einflusses der Kommerzialisierung des Gemüseanbaus auf das pro-Kopf Einkommen bestätigt, dass arme Haushalte von der Kommerzialisierung profitieren können. Ein zweistufiges Fixed-Effekt-Modell, das sowohl beobachtbare als auch unbeobachtbare Heterogenität berücksichtigt, zeigt, dass Haushalte, die Gemüse für den Exportmarkt produzieren, ein um 39% höheres Einkommen haben als Haushalte, die nicht für diesen Markt produzieren. Die Untersuchung der Rolle beider Geschlechter bei der Produktion von Gemüse ergibt, dass Frauen der Zugang zu Land, Wasser, landwirtschaftlichen Krediten sowie Produktions- und Marktinformationen verwehrt wird. Die empirische Analyse zeigt aber auch, dass Verbesserungen in der Produktivität des Bodens und die Unterstützung von Frauen beim Zugang zu landwirtschaftlichen Trainings und Beratung ihre Beteiligung am Exportmarkt erhöhen kann. Die Evaluierung der vertraglichen Bedingungen zwischen kleinbäuerlichen Gemüseproduzenten und Exporteuren beweist, dass die Governance-Strukturen verschiedener vertraglicher Arrangements großen Einfluss auf die Transaktionen zwischen beiden Parteien haben. Die verschiedenen institutionellen Arrangements beeinflussen die Profitabilität und variieren zwischen den Regionen. Außerdem zeigt eine dynamische Cost-Benefit Analyse des Anbaus von Gartenbohnen, dass individuelle Verträge finanziell profitabler sind als Gruppenverträge, unverbindliche Verträge oder direkter Verkauf.

Interventionen, die die Kommerzialisierung von Kleinbauern fördern könnten, beinhalten z.B. die Verbesserung der physischen Infrastruktur und die Bereitstellung von günstigen Krediten, die den Kauf von Inputs wie Saatgut und Dünger ermöglichen. Außerdem sollten bei der Schaffung von Wertschöpfungsketten Frauen besonders geschult werden. Kollektives Handeln und vertragliche Bedingungen zwischen Kleinbauern und Händlern sollten gestärkt werden, unter Berücksichtigung der Rolle von Vermittlern, Zwischenhändlern und saisonalen Exporteuren.

Development Economics and Policy

Series edited by Franz Heidhues, Joachim von Braun and Manfred Zeller

Band 1 Andrea Fadani: Agricultural Price Policy and Export and Food Production in Cameroon. A Farming Systems Analysis of Pricing Policies. The Case of Coffee-Based Farming Systems. 1999.

Band 2 Heike Michelsen: Auswirkungen der Währungsunion auf den Strukturanpassungsprozeß der Länder der afrikanischen Franc-Zone. 1995.

Band 3 Stephan Bea: Direktinvestitionen in Entwicklungsländern. Auswirkungen von Stabilisierungsmaßnahmen und Strukturreformen in Mexiko. 1995.

Band 4 Franz Heidhues / François Kamajou: Agricultural Policy Analysis – Proceedings of an International Seminar, held at the University of Dschang, Cameroon on May 26 and 27 1994, funded by the European Union under the Science and Technology Program (STD). 1996.

Band 5 Elke M. Förster: Protection or Liberalization? A Policy Analysis of the Korean Beef Sector. 1996.

Band 6 Gertrud Schrieder: The Role of Rural Finance for Food Security of the Poor in Cameroon. 1996.

Band 7 Nestor R. Ahoyo Adjovi: Economie des Systèmes de Production intégrant la Culture de Riz au Sud du Bénin: Potentialités, Contraintes et Perspectives. 1996.

Band 8 Jenny Müller: Income Distribution in the Agricultural Sector of Thailand. Empirical Analysis and Policy Options. 1996.

Band 9 Michael Brüntrup: Agricultural Price Policy and its Impact on Production, Income, Employment and the Adoption of Innovations. A Farming Systems Based Analysis of Cotton Policy in Northern Benin. 1997.

Band 10 Justin Bomda: Déterminants de l'Epargne et du Crédit, et leurs Implications pour le Développement du Système Financier Rural au Cameroun. 1998.

Band 11 John M. Msuya: Nutrition Improvement Projects in Tanzania: Implementation, Determinants of Performance, and Policy Implications. 1998.

Band 12 Andreas Neef: Auswirkungen von Bodenrechtswandel auf Ressourcennutzung und wirtschaftliches Verhalten von Kleinbauern in Niger und Benin. 1999.

Band 13 Susanna Wolf (ed.): The Future of EU-ACP Relations. 1999.

Band 14 Franz Heidhues / Gertrud Schrieder (eds.): Romania – Rural Finance in Transition Economies. 2000.

Band 15 Katinka Weinberger: Women's Participation. An Economic Analysis in Rural Chad and Pakistan. 2000.

Band 16 Christof Batzlen: Migration and Economic Development. Remittances and Investments in South Asia: A Case Study of Pakistan. 2000.

Band 17 Matin Qaim: Potential Impacts of Crop Biotechnology in Developing Countries. 2000.

Band 18 Jean Senahoun: Programmes d'ajustement structurel, sécurité alimentaire et durabilité agricole. Une approche d'analyse intégrée, appliquée au Bénin. 2001.

Band 19 Torsten Feldbrügge: Economics of Emergency Relief Management in Developing Countries. With Case Studies on Food Relief in Angola and Mozambique. 2001.

Band 20 Claudia Ringler: Optimal Allocation and Use of Water Resources in the Mekong River Basin: Multi-Country and Intersectoral Analyses. 2001.

Band 21 Arnim Kuhn: Handelskosten und regionale (Des-)Integration. Russlands Agrarmärkte in der Transformation. 2001.

Band 22 Ortrun Anne Gronski: Stock Markets and Economic Growth. Evidence from South Africa. 2001.

Band 23 Patrick Webb / Katinka Weinberger (eds.): Women Farmers. Enhancing Rights, Recognition and Productivity. 2001.

Band 24 Mingzhi Sheng: Lebensmittelkonsum und -konsumtrends in China. Eine empirische Analyse auf der Basis ökonometrischer Nachfragemodelle. 2002.

Band 25 Maria Iskandarani: Economics of Household Water Security in Jordan. 2002.

Band 26 Romeo Bertolini: Telecommunication Services in Sub-Saharan Africa. An Analysis of Access and Use in the Southern Volta Region in Ghana. 2002.

Band 27 Dietrich Müller-Falcke: Use and Impact of Information and Communication Technologies in Developing Countries' Small Businesses. Evidence from Indian Small Scale Industry. 2002.

Band 28 Wolfram Erhardt: Financial Markets for Small Enterprises in Urban and Rural Northern Thailand. Empirical Analysis on the Demand for and Supply of Financial Services, with Particular Emphasis on the Determinants of Credit Access and Borrower Transaction Costs. 2002.

Band 29 Wensheng Wang: The Impact of Information and Communication Technologies on Farm Households in China. 2002.

Band 30 Shyamal K. Chowdhury: Institutional and Welfare Aspects of the Provision and Use of Information and Communication Technologies in the Rural Areas of Bangladesh and Peru. 2002.

Band 31 Annette Luibrand: Transition in Vietnam. Impact of the Rural Reform Process on an Ethnic Minority. 2002.

Band 32 Felix Ankomah Asante: Economic Analysis of Decentralisation in Rural Ghana. 2003.

Band 33 Chodechai Suwanaporn: Determinants of Bank Lending in Thailand: An Empirical Examination for the Years 1992 to 1996. 2003.

Band 34 Abay Asfaw: Costs of Illness, Demand for Medical Care, and the Prospect of Community Health Insurance Schemes in the Rural Areas of Ethiopia. 2003.

Band 35 Gi-Soon Song: The Impact of Information and Communication Technologies (ICTs) on Rural Households. A Holistic Approach Applied to the Case of Lao People's Democratic Re- public. 2003.

Band 36 Daniela Lohlein: An Economic Analysis of Public Good Provision in Rural Russia. The Case of Education and Health Care. 2003.

Band 37 Johannes Woelcke. Bio-Economics of Sustainable Land Management in Uganda. 2003.

Band 38 Susanne M. Ziemek: The Economics of Volunteer Labor Supply. An Application to Countries of a Different Development Level. 2003.

Band 39 Doris Wiesmann: An International Nutrition Index. Concept and Analyses of Food Insecurity and Undernutrition at Country Levels. 2004.

Band 40 Isaac Osei-Akoto: The Economics of Rural Health Insurance. The Effects of Formal and Informal Risk-Sharing Schemes in Ghana. 2004.

Band 41 Yuansheng Jiang: Health Insurance Demand and Health Risk Management in Rural China. 2004.

Band 42 Roukayatou Zimmermann: Biotechnology and Value-added Traits in Food Crops: Relevance for Developing Countries and Economic Analyses. 2004.

Band 43 F. Markus Kaiser: Incentives in Community-based Health Insurance Schemes. 2004.

Band 44 Thomas Herzfeld: *Corruption begets Corruption.* Zur Dynamik und Persistenz der Korruption. 2004.

Band 45 Edilegnaw Wale Zegeye: The Economics of On-Farm Conservation of Crop Diversity in Ethiopia: Incentives, Attribute Preferences and Opportunity Costs of Maintaining Local Varieties of Crops. 2004.

Band 46 Adama Konseiga: Regional Integration Beyond the Traditional Trade Benefits: Labor Mobility contribution. The Case of Burkina Faso and Côte d'Ivoire. 2005.

Band 47 Beyene Tadesse Ferenji: The Impact of Policy Reform and Institutional Transformation on Agricultural Performance. An Economic Study of Ethiopian Agriculture. 2005.

Band 48 Sabine Daude: Agricultural Trade Liberalization in the WTO and Its Poverty Implications. A Study of Rural Households in Northern Vietnam. 2005.

Band 49 Kadir Osman Gyasi: Determinants of Success of Collective Action on Local Commons. An Empirical Analysis of Community-Based Irrigation Management in Northern Ghana. 2005.

Band 50 Borbala E. Balint: Determinants of Commercial Orientation and Sustainability of Agricultural Production of the Individual Farms in Romania. 2006.

Band 51 Pamela Marinda: Effects of Gender Inequality in Resource Ownership and Access on Household Welfare and Food Security in Kenya. A Case Study of West Pokot District. 2006.

Band 52 Charles Palmer: The Outcomes and their Determinants from Community-Company Contracting over Forest Use in Post-Decentralization Indonesia. 2006.

Band 53 Hardwick Tchale: Agricultural Policy and Soil Fertility Management in the Maize-based Smallholder Farming System in Malawi. 2006.

Band 54 John Kedi Mduma: Rural Off-Farm Employment and its Effects on Adoption of Labor Intensive Soil Conserving Measures in Tanzania. 2006.

Band 55 Mareike Meyn: The Impact of EU Free Trade Agreements on Economic Development and Regional Integration in Southern Africa. The Example of EU-SACU Trade Relations. 2006.

Band 56 Clemens Breisinger: Modelling Infrastructure Investments, Growth and Poverty Impact. A Two-Region Computable General Equilibrium Perspective on Vietnam. 2006.

Band 57 Meike Wollni: Coping with the Coffee Crisis. An Analysis of the Production and Marketing Performance of Coffee Farmers in Costa Rica. 2007.

Band 58 Franklin Simtowe: Performance and Impact of Microfinance. Evidence from Joint Liability Lending Programs in Malawi. 2008.

Band 59 Xiangping Jia: Credit Rationing and Institutional Constraint. Evidence from Rural China. 2008.

Band 60 Holger Seebens: The Economics of Gender and the Household in Developing Countries. 2009.

Band 61 Stephan Piotrowski: Land Property Rights and Natural Resource Use. An Analysis of Household Behavior in Rural China. 2009.

Band 62 Sebastian M. Scholz: Rural Development through Carbon Finance. Forestry Projects under the Clean Development Mechanism of the Kyoto Protocol. Assessing Smallholder Participation by Structural Equation Modeling. 2009.

Band 63 Jakob Rupert Friederichsen: Opening Up Knowledge Production through Participatory Research? Agricultural Research for Vietnam's Northern Uplands. 2009.

Band 64 Olivier Ecker: Economics of Micronutrient Malnutrition. The Demand for Nutrients in Sub-Saharan Africa. 2009.

Band 65 Julia Johannsen: Operational Assessment of Monetary Poverty by Proxy Means Tests. 2009

Band 66 Ephraim Nkonya / Nicolas Gerber / Philipp Baumgartner / Joachim von Braun / Alessandro De Pinto / Valerie Graw / Edward Kato / Julia Kloos / Teresa Walter: The Economics of Land Degradation. Toward an Integrated Global Assessment. 2011.

Band 67 S. Idriss Nazaire Houssou: Operational Poverty Targeting by Proxy Means Tests. Models and Policy Simulations for Malawi. 2013.

Band 68 Abdul Salam Lodhi: Education, Child Labor and Human Capital Formation in Selected Urban and Rural Settings of Pakistan. 2013.

Band 69 Evita Hanie Pangaribowo: Household Food Consumption, Women´s Asset and Food Policy in Indonesia. 2013.

Band 70 Dan Liu: China's New Rural Cooperative Medical Scheme. Evolution, Design and Impacts. 2013.

Band 71 Camille Saint-Macary: Microeconomic Impacts of Institutional Change in Vietnam's Northern Uplands. Empirical Studies on Social Capital, Land and Credit Institutions. 2014.

Band 72 Beatrice Wambui Muriithi: Commercialization of Smallholder Horticultural Farming in Kenya. Poverty, Gender, and Institutional Arrangements. 2014.

www.peterlang.com